Praise for Michio Kaku's

PHYSICS OF THE IMPOSSIBLE

"Kaku encourages us to take seriously ideas the world's great intellects consider crazy, reminding us that these same powerful minds sometimes wonder whether such way-out theories and models of the universe are crazy enough to be true." —*The Seattle Times*

"An invigorating experience." —*The Christian Science Monitor*

"A stimulating and entertaining read, underlining the need for know-all scientists to avoid smug complacency." —*Physics World*

"Enlightening and easy to read. . . . Kaku is cheerfully optimistic about the wonders in store for us." —*The Spectator*

"Kaku is an effective and gifted dramatiser of highly complex ideas. If you want to know what the implications would be of room-temperature superconductors, or all about tachyons, particles that travel faster than the speed of light and pass through all points of the universe simultaneously, then this is the place to find out."

—*The Sunday Times* (London)

"[Kaku] has a knack for making complex ideas entertaining."

—*The Charlotte Observer*

"Science and science fiction buffs can easily follow Kaku's explanations as he shows that in the wonderful worlds of science, impossible things are happening every day." —*Publishers Weekly*

"[A] marvellously approachable book. . . . Kaku tackles with absolute seriousness the questions that excite every small boy."

—*The Sunday Telegraph* (London)

Michio Kaku

PHYSICS OF THE IMPOSSIBLE

Michio Kaku is the Henry Semat Professor of Theoretical Physics at the Graduate Center of the City University of New York. He is the cofounder of string field theory. He has written several books, including *Parallel Worlds* and *Beyond Einstein*, and his bestseller, *Hyperspace*, was voted one of the best science books of the year by *The New York Times* and *The Washington Post*. Dr. Kaku has hosted numerous science documentaries for the Discovery Channel and BBC-TV, is a frequent guest on national TV, and his nationally syndicated radio program is heard in 130 cities. He lives in New York City.

www.mkaku.org

Also by Michio Kaku

PARALLEL WORLDS

EINSTEIN'S COSMOS

VISIONS

HYPERSPACE

BEYOND EINSTEIN

PHYSICS OF THE IMPOSSIBLE

A Scientific Exploration into the World of Phasers, Force Fields, Teleportation, and Time Travel

Michio Kaku

Anchor Books
A Division of Random House, Inc.
New York

To my loving wife, Shizue,

and to

Michelle and Alyson

FIRST ANCHOR BOOKS EDITION, APRIL 2009

The Library of Congress has cataloged the Doubleday edition as follows:
Kaku, Michio.
Physics of the impossible : a scientific exploration into the world of phasers, force fields, teleportation, and time travel / Michio Kaku.–1st ed.
p. cm.
Includes bibliographical references and index.
1. Physics–Miscellanea. 2. Science–Miscellanea. 3. Mathematical physics–Miscellanea. 4. Physics in literature. 5. Human-machine systems. I. Title.
QC75.K18 2008
530–dc25 2007030290

Anchor ISBN: 978-0-307-27882-1

www.anchorbooks.com

Printed in the United States of America
10 9 8 7 6 5 4 3 2 1

CONTENTS

PREFACE

If at first an idea does not sound absurd,

then there is no hope for it.

–ALBERT EINSTEIN

One day, would it be possible to walk through walls? To build starships that can travel faster than the speed of light? To read other people's minds? To become invisible? To move objects with the power of our minds? To transport our bodies instantly through outer space?

Since I was a child, I've always been fascinated by these questions. Like many physicists, when I was growing up, I was mesmerized by the possibility of time travel, ray guns, force fields, parallel universes, and the like. Magic, fantasy, science fiction were all a gigantic playground for my imagination. They began my lifelong love affair with the impossible.

I remember watching the old *Flash Gordon* reruns on TV. Every Saturday, I was glued to the TV set, marveling at the adventures of Flash, Dr. Zarkov, and Dale Arden and their dazzling array of futuristic technology: the rocket ships, invisibility shields, ray guns, and cities in the sky. I never missed a week. The program opened up an entirely new world for me. I was thrilled by the thought of one day rocketing to an alien planet and exploring its strange terrain. Being pulled into the orbit of these fantastic inventions I knew that my own destiny was somehow wrapped up with the marvels of the science that the show promised.

As it turns out, I was not alone. Many highly accomplished scientists originally became interested in science through exposure to science fiction. The great astronomer Edwin Hubble was fascinated by the works of Jules Verne. As a result of reading Verne's work, Hubble abandoned a promising career in law, and, disobeying his father's wishes, set off on a career in science. He eventually became the greatest astronomer of the twentieth century. Carl Sagan, noted astronomer and bestselling author, found his imagination set afire by reading Edgar Rice Burroughs's John Carter of Mars novels. Like John Carter, he dreamed of one day exploring the sands of Mars.

I was just a child the day when Albert Einstein died, but I remember people talking about his life, and death, in hushed tones. The next day I saw in the newspapers a picture of his desk, with the unfinished manuscript of his greatest, unfinished work. I asked myself, What could be so important that the greatest scientist of our time could not finish it? The article claimed that Einstein had an impossible dream, a problem so difficult that it was not possible for a mortal to finish it. It took me years to find out what that manuscript was about: a grand, unifying "theory of everything." His dream–which consumed the last three decades of his life–helped me to focus my own imagination. I wanted, in some small way, to be part of the effort to complete Einstein's work, to unify the laws of physics into a single theory.

As I grew older I began to realize that although Flash Gordon was the hero and always got the girl, it was the scientist who actually made the TV series work. Without Dr. Zarkov, there would be no rocket ship, no trips to Mongo, no saving Earth. Heroics aside, without science there is no science fiction.

I came to realize that these tales were simply impossible in terms of the science involved, just flights of the imagination. Growing up meant putting away such fantasy. In real life, I was told, one had to abandon the impossible and embrace the practical.

However, I concluded that if I was to continue my fascination with the impossible, the key was through the realm of physics. Without a solid background in advanced physics, I would be forever speculating about futuristic technologies without understanding whether or not they were possible. I realized I needed to immerse myself in advanced mathematics and learn theoretical physics. So that is what I did.

In high school for my science fair project I assembled an atom smasher in my mom's garage. I went to the Westinghouse company and gathered 400 pounds of scrap transformer steel. Over Christmas I wound 22 miles of copper wire on the high school football field. Eventually I built a 2.3-million-electron-volt betatron particle accelerator, which consumed 6 kilowatts of power (the entire output of my house) and generated a magnetic field of 20,000 times the Earth's magnetic field. The goal was to generate a beam of gamma rays powerful enough to create antimatter.

My science fair project took me to the National Science Fair and eventually fulfilled my dream, winning a scholarship to Harvard, where I could finally pursue my goal of becoming a theoretical physicist and follow in the footsteps of my role model, Albert Einstein.

Today I receive e-mails from science fiction writers and screenwriters asking me to help them sharpen their own tales by exploring the limits of the laws of physics.

The "Impossible" Is Relative

As a physicist, I have learned that the "impossible" is often a relative term. Growing up, I remember my teacher one day walking up to the map of the Earth on the wall and pointing out the coastlines of South America and Africa. Wasn't it an odd coincidence, she said, that the two coastlines fit together, almost like a jigsaw puzzle? Some scientists, she said, speculated that perhaps they were once part of the same, vast continent. But that was silly. No force could possibly push two gigantic continents apart. Such thinking was impossible, she concluded.

Later that year we studied the dinosaurs. Wasn't it strange, our teacher told us, that the dinosaurs dominated the Earth for millions of years, and then one day they all vanished? No one knew why they had all died off. Some paleontologists thought that maybe a meteor from space had killed them, but that was impossible, more in the realm of science fiction.

Today we now know that through plate tectonics the continents do move, and that 65 million years ago a gigantic meteor measuring six miles across most likely did obliterate the dinosaurs and much of life on Earth. In my own short lifetime I have seen the seemingly impossible become established scientific fact over and over again. So is it impossible to think we might one day be able to teleport ourselves from one place to another, or build a spaceship that will one day take us light-years away to the stars?

Normally such feats would be considered impossible by today's physicists. Might they become possible within a few centuries? Or in ten thousand years, when our technology is more advanced? Or in a million years? To put it another way, if we were to somehow encounter a civilization a million years more advanced than ours, would their everyday technology appear to be "magic" to us? That, at its heart, is one of the central questions running through this book; just because something is "impossible" today, will it remain impossible centuries or millions of years into the future?

Given the remarkable advances in science in the past century, especially the creation of the quantum theory and general relativity, it is now possible to give rough estimates of when, if ever, some of these fantastic technologies may be realized. With the coming of even more advanced theories, such as string theory, even concepts bordering on science fiction, such as time travel and parallel universes, are now being re-evaluated by physicists. Think back 150 years to those technological advances that were declared "impossible" by scientists at the time and that have now become part of our everyday lives. Jules Verne wrote a novel in 1863, *Paris in the Twentieth Century,* which was locked away and forgotten for over a century until it was accidentally discovered by his great-grandson and published for the first time in 1994. In it Verne predicted what Paris might look like in the year 1960. His novel was filled with technology that was clearly considered impossible in the nineteenth century, including fax machines, a worldwide communications network, glass skyscrapers, gas-powered automobiles, and high-speed elevated trains.

Not surprisingly, Verne could make such stunningly accurate predictions because he was immersed in the world of science, picking the brains of scientists around him. A deep appreciation for the fundamentals of science allowed him to make such startling predictions.

Sadly, some of the greatest scientists of the nineteenth century took the opposite position and declared any number of technologies to be hopelessly impossible. Lord Kelvin, perhaps the most prominent physicist of the Victorian era (he is buried next to Isaac Newton in Westminster Abbey), declared that "heavier than air" devices such as the airplane were impossible. He thought X-rays were a hoax and that radio had no future. Lord Rutherford, who discovered the nucleus of the atom, dismissed the possibility of building an atomic bomb, comparing it to "moonshine." Chemists of the nineteenth century declared the search for the philosopher's stone, a fabled substance that can turn lead into gold, a scientific dead end. Nineteenth-century chemistry was based on the fundamental immutability of the elements, like lead. Yet with today's atom smashers, we can, in principle, turn lead atoms into gold. Think how fantastic today's televisions, computers, and Internet would have seemed at the turn of the twentieth century.

More recently, black holes were once considered to be science fiction. Einstein himself wrote a paper in 1939 that "proved" that black holes could never form. Yet today the Hubble Space Telescope and the Chandra X-ray telescope have revealed thousands of black holes in space.

The reason that these technologies were deemed "impossibilities" is that the basic laws of physics and science were not known in the nineteenth century and the early part of the twentieth. Given the huge gaps in the understanding of science at the time, especially at the atomic level, it's no wonder such advances were considered impossible.

Studying the Impossible

Ironically, the serious study of the impossible has frequently opened up rich and entirely unexpected domains of science. For example, over the centuries the frustrating and futile search for a "perpetual motion machine" led physicists to conclude that such a machine was impossible, forcing them to postulate the conservation of energy and the three laws of thermodynamics. Thus the futile search to build perpetual motion machines helped to open up the entirely new field of thermodynamics, which in part laid the foundation of the steam engine, the machine age, and modern industrial society.

At the end of the nineteenth century, scientists decided that it was "impossible" for the Earth to be billions of years old. Lord Kelvin declared flatly that a molten Earth would cool down in 20 to 40 million years, contradicting the geologists and Darwinian biologists who claimed that the Earth might be billions of years old. The impossible was finally proven to be possible with the discovery of the nuclear force by Madame Curie and others, showing how the center of the Earth, heated by radioactive decay, could indeed be kept molten for billions of years.

We ignore the impossible at our peril. In the 1920s and 1930s Robert Goddard, the founder of modern rocketry, was the subject of intense criticism by those who thought that rockets could never travel in outer space. They sarcastically called his pursuit Goddard's Folly. In 1921 the editors of the *New York Times* railed against Dr. Goddard's work: "Professor Goddard does not know the relation between action and reaction and the need to have something better than a vacuum against which to react. He seems to lack the basic knowledge ladled out daily in high schools." Rockets were impossible, the editors huffed, because there was no air to push against in outer space. Sadly, one head of state did understand the implications of Goddard's "impossible" rockets–Adolf Hitler. During World War II, Germany's barrage of impossibly advanced V-2 rockets rained death and destruction on London, almost bringing it to its knees.

Studying the impossible may have also changed the course of world history. In the 1930s it was widely believed, even by Einstein, that an atomic bomb was "impossible." Physicists knew that there was a tremendous amount of energy locked deep inside the atom's nucleus, according to Einstein's equation $E = mc^2$, but the energy released by a single nucleus was too insignificant to consider. But atomic physicist Leo Szilard remembered reading the 1914 H. G. Wells novel, *The World Set Free*, in which Wells predicted the development of the atomic bomb. In the book he stated that the secret of the atomic bomb would be solved by a physicist in 1933. By chance Szilard stumbled upon this book in 1932. Spurred on by the novel, in 1933, precisely as predicted by Wells some two decades earlier, he hit upon the idea of magnifying the power of a single atom via a chain reaction, so that the energy of splitting a single uranium nucleus could be magnified by many trillions. Szilard then set into motion a series of key experiments and secret negotiations between Einstein and President Franklin Roosevelt that would lead to the Manhattan Project, which built the atomic bomb.

Time and again we see that the study of the impossible has opened up entirely new vistas, pushing the boundaries of physics and chemistry and forcing scientists to redefine what they mean by "impossible." As Sir William Osler once said, "The philosophies of one age have become the absurdities of the next, and the foolishness of yesterday has become the wisdom of tomorrow."

Many physicists subscribe to the famous dictum of T. H. White, who wrote in *The Once and Future King*, "Anything that is not forbidden, is mandatory!" In physics we find evidence of this all the time. Unless there is a law of physics explicitly preventing a new phenomenon, we eventually find that it exists. (This has happened several times in the search for new subatomic particles. By probing the limits of what is forbidden, physicists have often unexpectedly discovered new laws of physics.) A corollary to T. H. White's statement might well be, "Anything that is not impossible, is mandatory!"

For example, cosmologist Stephen Hawking tried to prove that time travel was impossible by finding a new law of physics that would forbid it, which he called the "chronology protection conjecture." Unfortunately, after many years of hard work he was unable to prove this principle. In fact, to the contrary, physicists have now demonstrated that a law that prevents time travel is beyond our present-day mathematics. Today, because there is no law of physics preventing the existence of time machines, physicists have had to take their possibility very seriously.

The purpose of this book is to consider what technologies are considered "impossible" today that might well become commonplace decades to centuries down the road.

Already one "impossible" technology is now proving to be possible: the notion of teleportation (at least at the level of atoms). Even a few years ago physicists would have said that sending or beaming an object from one point to another violated the laws of quantum physics. The writers of the original *Star Trek* television series, in fact, were so stung by the criticism from physicists that they added "Heisenberg compensators" to explain their teleporters in order to address this flaw. Today, because of a recent breakthrough, physicists can teleport atoms across a room or photons under the Danube River.

Predicting the Future

It is always a bit dangerous to make predictions, especially ones set centuries to thousands of years in the future. The physicist Niels Bohr was fond of saying, "Prediction is very hard to do. Especially about the future." But there is a fundamental difference between the time of Jules Verne and the present. Today the fundamental laws of physics are basically understood. Physicists today understand the basic laws extending over a staggering forty-three orders of magnitude, from the interior of the proton out to the expanding universe. As a result, physicists can state, with reasonable confidence, what the broad outlines of future technology might look like, and better differentiate between those technologies that are merely improbable and those that are truly impossible.

In this book, therefore, I divide the things that are "impossible" into three categories.

The first are what I call *Class I impossibilities.* These are technologies that are impossible today but that do not violate the known laws of physics. So they might be possible in this century, or perhaps the next, in modified form. They include teleportation, antimatter engines, certain forms of telepathy, psychokinesis, and invisibility.

The second category is what I term *Class II impossibilities.* These are technologies that sit at the very edge of our understanding of the physical world. If they are possible at all, they might be realized on a scale of millennia to millions of years in the future. They include time machines, the possibility of hyperspace travel, and travel through wormholes.

The final category is what I call *Class III impossibilities.* These are technologies that violate the known laws of physics. Surprisingly, there are very few such impossible technologies. If they do turn out to be possible, they would represent a fundamental shift in our understanding of physics.

This classification is significant, I feel, because so many technologies in science fiction are dismissed by scientists as being totally impossible, when what they actually mean is that they are impossible for a primitive civilization like ours. Alien visitations, for example, are usually considered impossible because the distances between the stars are so vast. While interstellar travel for our civilization is clearly impossible, it may be possible for a civilization centuries to thousands or millions of years ahead of ours. So it is important to rank such "impossibilities." Technologies that are impossible for our current civilization are not necessarily impossible for other types of civilizations. Statements about what is possible and impossible have to take into account technologies that are millennia to millions of years ahead of ours.

Carl Sagan once wrote, "What does it mean for a civilization to be a million years old? We have had radio telescopes and spaceships for a few decades; our technical civilization is a few hundred years old . . . an advanced civilization millions of years old is as much beyond us as we are beyond a bush baby or a macaque."

In my own research I focus professionally on trying to complete Einstein's dream of a "theory of everything." Personally, I find it quite exhilarating to work on a "final theory" that may ultimately answer some of the most difficult "impossible" questions in science today, such as whether time travel is possible, what lies at the center of a black hole, or what happened before the big bang. I still daydream about my lifelong love affair with the impossible, and wonder when and if some of these impossibilities might enter the ranks of the everyday.

ACKNOWLEDGMENTS

The material in this book ranges over many fields and disciplines, as well as the work of many outstanding scientists. I would like to thank the following individuals, who have graciously given their time for lengthy interviews, consultations, and interesting, stimulating conversations:

Leon Lederman, Nobel laureate, Illinois Institute of Technology
Murray Gell-Mann, Nobel laureate, Santa Fe Institute and Cal Tech
The late Henry Kendall, Nobel laureate, MIT
Steven Weinberg, Nobel laureate, University of Texas at Austin
David Gross, Nobel laureate, Kavli Institute for Theoretical Physics
Frank Wilczek, Nobel laureate, MIT
Joseph Rotblat, Nobel laureate, St. Bartholomew's Hospital
Walter Gilbert, Nobel laureate, Harvard University
Gerald Edelman, Nobel laureate, Scripps Research Institute
Peter Doherty, Nobel laureate, St. Jude Children's Research Hospital
Jared Diamond, Pulitzer Prize winner, UCLA
Stan Lee, creator of Marvel Comics and Spiderman
Brian Greene, Columbia University, author of *The Elegant Universe*
Lisa Randall, Harvard University, author of *Warped Passages*
Lawrence Krauss, Case Western University, author of *The Physics of Star Trek*
J. Richard Gott III, Princeton University, author of *Time Travel in Einstein's Universe*
Alan Guth, physicist, MIT, author of *The Inflationary Universe*

John Barrow, physicist, Cambridge University, author of *Impossibility*
Paul Davies, physicist, author of *Superforce*
Leonard Susskind, physicist, Stanford University
Joseph Lykken, physicist, Fermi National Laboratory
Marvin Minsky, MIT, author of *The Society of Minds*
Ray Kurzweil, inventor, author of *The Age of Spiritual Machines*
Rodney Brooks, director of MIT Artificial Intelligence Laboratory
Hans Moravec, author of *Robot*
Ken Croswell, astronomer, author of *Magnificent Universe*
Don Goldsmith, astronomer, author of *Runaway Universe*
Neil de Grasse Tyson, director of Hayden Planetarium, New York City
Robert Kirshner, astronomer, Harvard University
Fulvia Melia, astronomer, University of Arizona
Sir Martin Rees, Cambridge University, author of *Before the Beginning*
Michael Brown, astronomer, Cal Tech
Paul Gilster, author of *Centauri Dreams*
Michael Lemonick, senior science editor of *Time* magazine
Timothy Ferris, University of California, author of *Coming of Age in the Milky Way*
The late Ted Taylor, designer of U.S. nuclear warheads
Freeman Dyson, Institute for Advanced Study, Princeton
John Horgan, Stevens Institute of Technology, author of *The End of Science*
The late Carl Sagan, Cornell University, author of *Cosmos*
Ann Druyan, widow of Carl Sagan, Cosmos Studios
Peter Schwarz, futurist, founder of Global Business Network
Alvin Toffler, futurist, author of *The Third Wave*
David Goodstein, assistant provost of Cal Tech
Seth Lloyd, MIT, author of *Programming the Universe*
Fred Watson, astronomer, author of *Star Gazer*
Simon Singh, author of *The Big Bang*
Seth Shostak, SETI Institute
George Johnson, *New York Times* science journalist
Jeffrey Hoffman, MIT, NASA astronaut

Tom Jones, NASA astronaut
Alan Lightman, MIT, author of *Einstein's Dreams*
Robert Zubrin, founder of Mars Society
Donna Shirley, NASA Mars program
John Pike, GlobalSecurity.org
Paul Saffo, futurist, Institute of the Future
Louis Friedman, cofounder of the Planetary Society
Daniel Werthheimer, SETI@home, University of California at Berkeley
Robert Zimmerman, author of *Leaving Earth*
Marcia Bartusiak, author of *Einstein's Unfinished Symphony*
Michael H. Salamon, NASA's Beyond Einstein program
Geoff Andersen, U.S. Air Force Academy, author of *The Telescope*

I would also like to thank my agent, Stuart Krichevsky, who has been at my side all these years, shepherding all my books, and also my editor, Roger Scholl, whose firm hand, sound judgment, and editorial experience have guided so many of my books. I would also like to thank my collegues at the City College of New York and the Graduate Center of the City University of New York, especially V. P. Nair and Dan Greenberger, who generously donated their time for discussions.

Part I

CLASS I IMPOSSIBILITIES

1: FORCE FIELDS

I. When a distinguished but elderly scientist states that something is possible, he is almost certainly right. When he states that something is impossible, he is very probably wrong.

II. The only way of discovering the limits of the possible is to venture a little way past them into the impossible.

III. Any sufficiently advanced technology is indistinguishable from magic.

—ARTHUR C. CLARKE'S THREE LAWS

"Shields up!"

In countless *Star Trek* episodes this is the first order that Captain Kirk barks out to the crew, raising the force fields to protect the starship *Enterprise* against enemy fire.

So vital are force fields in *Star Trek* that the tide of the battle can be measured by how the force field is holding up. Whenever power is drained from the force fields, the *Enterprise* suffers more and more damaging blows to its hull, until finally surrender is inevitable.

So what is a force field? In science fiction it's deceptively simple: a thin, invisible yet impenetrable barrier able to deflect lasers and rockets alike. At first glance a force field looks so easy that its creation as a battlefield shield seems imminent. One expects that any day some enterprising inventor will announce the discovery of a defensive force field. But the truth is far more complicated.

In the same way that Edison's lightbulb revolutionized modern civilization, a force field could profoundly affect every aspect of our lives. The military could use force fields to become invulnerable, creating an impenetrable shield against enemy missiles and bullets. Bridges, superhighways, and roads could in theory be built by simply pressing a button. Entire cities could sprout instantly in the desert, with skyscrapers made entirely of force fields. Force fields erected over cities could enable their inhabitants to modify the effects of their weather–high winds, blizzards, tornados–at will. Cities could be built under the oceans within the safe canopy of a force field. Glass, steel, and mortar could be entirely replaced.

Yet oddly enough a force field is perhaps one of the most difficult devices to create in the laboratory. In fact, some physicists believe it might actually be impossible, without modifying its properties.

MICHAEL FARADAY

The concept of force fields originates from the work of the great nineteenth-century British scientist Michael Faraday.

Faraday was born to working-class parents (his father was a blacksmith) and eked out a meager existence as an apprentice bookbinder in the early 1800s. The young Faraday was fascinated by the enormous breakthroughs in uncovering the mysterious properties of two new forces: electricity and magnetism. Faraday devoured all he could concerning these topics and attended lectures by Professor Humphrey Davy of the Royal Institution in London.

One day Professor Davy severely damaged his eyes in a chemical accident and hired Faraday to be his secretary. Faraday slowly began to win the confidence of the scientists at the Royal Institution and was allowed to conduct important experiments of his own, although he was often slighted. Over the years Professor Davy grew increasingly jealous of the brilliance shown by his young assistant, who was a rising star in experimental circles, eventually eclipsing Davy's own fame. After Davy died in 1829 Faraday was free to make a series of stunning breakthroughs that led to the creation of generators that would energize entire cities and change the course of world civilization.

The key to Faraday's greatest discoveries was his "force fields." If one places iron filings over a magnet, one finds that the iron filings create a spiderweb-like pattern that fills up all of space. These are Faraday's lines of force, which graphically describe how the force fields of electricity and magnetism permeate space. If one graphs the magnetic fields of the Earth, for example, one finds that the lines emanate from the north polar region and then fall back to the Earth in the south polar region. Similarly, if one were to graph the electric field lines of a lightning rod in a thunderstorm, one would find that the lines of force concentrate at the tip of the lightning rod. Empty space, to Faraday, was not empty at all, but was filled with lines of force that could make distant objects move. (Because of Faraday's poverty-stricken youth, he was illiterate in mathematics, and as a consequence his notebooks are full not of equations but of hand-drawn diagrams of these lines of force. Ironically, his lack of mathematical training led him to create the beautiful diagrams of lines of force that now can be found in any physics textbook. In science a physical picture is often more important than the mathematics used to describe it.)

Historians have speculated on how Faraday was led to his discovery of force fields, one of the most important concepts in all of science. In fact, the *sum total of all modern physics* is written in the language of Faraday's fields. In 1831, he made the key breakthrough regarding force fields that changed civilization forever. One day, he was moving a child's magnet over a coil of wire and he noticed that he was able to generate an electric current in the wire, without ever touching it. This meant that a magnet's invisible field could push electrons in a wire across empty space, creating a current.

Faraday's "force fields," which were previously thought to be useless, idle doodlings, were real, material forces that could move objects and generate power. Today the light that you are using to read this page is probably energized by Faraday's discovery about electromagnetism.

A spinning magnet creates a force field that pushes the electrons in a wire, causing them to move in an electrical current. This electricity in the wire can then be used to light up a lightbulb. This same principle is used to generate electricity to power the cities of the world. Water flowing across a dam, for example, causes a huge magnet in a turbine to spin, which then pushes the electrons in a wire, forming an electric current that is sent across high-voltage wires into our homes.

In other words, the force fields of Michael Faraday are the forces that drive modern civilization, from electric bulldozers to today's computers, Internet, and iPods.

Faraday's force fields have been an inspiration for physicists for a century and a half. Einstein was so inspired by them that he wrote his theory of gravity in terms of force fields. I, too, was inspired by Faraday's work. Years ago I successfully wrote the theory of strings in terms of the force fields of Faraday, thereby founding string field theory. In physics when someone says, "He thinks like a line of force," it is meant as a great compliment.

THE FOUR FORCES

Over the last two thousand years one of the crowning achievements of physics has been the isolation and identification of the four forces that rule the universe. All of them can be described in the language of fields introduced by Faraday. Unfortunately, however, none of them has quite the properties of the force fields described in most science fiction. These forces are

1. *Gravity*, the silent force that keeps our feet on the ground, prevents the Earth and the stars from disintegrating, and holds the solar system and galaxy together. Without gravity, we would be flung off the Earth into space at the rate of 1,000 miles per hour by the spinning planet. The problem is that gravity has precisely the opposite properties of a force field found in science fiction. Gravity is attractive, not repul-

sive; is extremely weak, relatively speaking; and works over enormous, astronomical distances. In other words, it is almost the opposite of the flat, thin, impenetrable barrier that one reads about in science fiction or one sees in science fiction movies. For example, it takes the entire planet Earth to attract a feather to the floor, but we can counteract Earth's gravity by lifting the feather with a finger. The action of our finger can counteract the gravity of an entire planet that weighs over six trillion trillion kilograms.

2. *Electromagnetism* (EM), the force that lights up our cities. Lasers, radio, TV, modern electronics, computers, the Internet, electricity, magnetism–all are consequences of the electromagnetic force. It is perhaps the most useful force ever harnessed by humans. Unlike gravity, it can be both attractive and repulsive. However, there are several reasons that it is unsuitable as a force field. First, it can be easily neutralized. Plastics and other insulators, for example, can easily penetrate a powerful electric or magnetic field. A piece of plastic thrown in a magnetic field would pass right through. Second, electromagnetism acts over large distances and cannot easily be focused onto a plane. The laws of the EM force are described by James Clerk Maxwell's equations, and these equations do not seem to admit force fields as solutions.

3 & 4. *The weak and strong nuclear forces.* The weak force is the force of radioactive decay. It is the force that heats up the center of the Earth, which is radioactive. It is the force behind volcanoes, earthquakes, and continental drift. The strong force holds the nucleus of the atom together. The energy of the sun and the stars originates from the nuclear force, which is responsible for lighting up the universe. The problem is that the nuclear force is a short-range force, acting mainly over the distance of a nucleus. Because it is so bound to the properties of nuclei, it is extremely hard to manipulate. At present the only ways we

have of manipulating this force are to blow subatomic particles apart in atom smashers or to detonate atomic bombs.

Although the force fields used in science fiction may not conform to the known laws of physics, there are still loopholes that might make the creation of such a force field possible. First, there may be a fifth force, still unseen in the laboratory. Such a force might, for example, work over a distance of only a few inches to feet, rather than over astronomical distances. (Initial attempts to measure the presence of such a fifth force, however, have yielded negative results.)

Second, it may be possible to use a plasma to mimic some of the properties of a force field. A plasma is the "fourth state of matter." Solids, liquids, and gases make up the three familiar states of matter, but the most common form of matter in the universe is plasma, a gas of ionized atoms. Because the atoms of a plasma are ripped apart, with electrons torn off the atom, the atoms are electrically charged and can be easily manipulated by electric and magnetic fields.

Plasmas are the most plentiful form of visible matter in the universe, making up the sun, the stars, and interstellar gas. Plasmas are not familiar to us because they are only rarely found on the Earth, but we can see them in the form of lightning bolts, the sun, and the interior of your plasma TV.

Plasma Windows

As noted above, if a gas is heated to a high enough temperature, thereby creating a plasma, it can be molded and shaped by magnetic and electrical fields. It can, for example, be shaped in the form of a sheet or window. Moreover, this "plasma window" can be used to separate a vacuum from ordinary air. In principle, one might be able to prevent the air within a spaceship from leaking out into space, thereby creating a convenient, transparent interface between outer space and the spaceship.

In the *Star Trek* TV series, such a force field is used to separate the shuttle bay, containing small shuttle craft, from the vacuum of outer space. Not only is it a clever way to save money on props, but it is a device that is possible.

The plasma window was invented by physicist Ady Herschcovitch in 1995 at the Brookhaven National Laboratory in Long Island, New York. He developed it to solve the problem of how to weld metals using electron beams. A welder's acetylene torch uses a blast of hot gas to melt and then weld metal pieces together. But a beam of electrons can weld metals faster, cleaner, and more cheaply than ordinary methods. The problem with electron beam welding, however, is that it needs to be done in a vacuum. This requirement is quite inconvenient, because it means creating a vacuum box that may be as big as an entire room.

Dr. Herschcovitch invented the plasma window to solve this problem. Only 3 feet high and less than 1 foot in diameter, the plasma window heats gas to 12,000°F, creating a plasma that is trapped by electric and magnetic fields. These particles exert pressure, as in any gas, which prevents air from rushing into the vacuum chamber, thus separating air from the vacuum. (When one uses argon gas in the plasma window, it glows blue, like the force field in *Star Trek.*)

The plasma window has wide applications for space travel and industry. Many times, manufacturing processes need a vacuum to perform microfabrication and dry etching for industrial purposes, but working in a vacuum can be expensive. But with the plasma window one can cheaply contain a vacuum with the flick of a button.

But can the plasma window also be used as an impenetrable shield? Can it withstand a blast from a cannon? In the future, one can imagine a plasma window of much greater power and temperature, sufficient to damage or vaporize incoming projectiles. But to create a more realistic force field, like that found in science fiction, one would need a combination of several technologies stacked in layers. Each layer might not be strong enough alone to stop a cannon ball, but the combination might suffice.

The outer layer could be a supercharged plasma window, heated to temperatures high enough to vaporize metals. A second layer could be a curtain of high-energy laser beams. This curtain, containing thousands of crisscrossing laser beams, would create a lattice that would heat up objects that passed through it, effectively vaporizing them. I will discuss lasers further in the next chapter.

And behind this laser curtain one might envision a lattice made of "carbon nanotubes," tiny tubes made of individual carbon atoms that are one atom thick and that are many times stronger than steel. Although the current world record for a carbon nanotube is only about 15 millimeters long, one can envision a day when we might be able to create carbon nanotubes of arbitrary length. Assuming that carbon nanotubes can be woven into a lattice, they could create a screen of enormous strength, capable of repelling most objects. The screen would be invisible, since each carbon nanotube is atomic in size, but the carbon nanotube lattice would be stronger than any ordinary material.

So, via a combination of plasma window, laser curtain, and carbon nanotube screen, one might imagine creating an invisible wall that would be nearly impenetrable by most means.

Yet even this multilayered shield would not completely fulfill all the properties of a science fiction force field–because it would be transparent and therefore incapable of stopping a laser beam. In a battle with laser cannons, the multilayered shield would be useless.

To stop a laser beam, the shield would also need to possess an advanced form of "photochromatics." This is the process used in sunglasses that darken by themselves upon exposure to UV radiation. Photochromatics are based on molecules that can exist in at least two states. In one state the molecule is transparent. But when it is exposed to UV radiation it instantly changes to the second form, which is opaque.

One day we might be able to use nanotechnology to produce a substance as tough as carbon nanotubes that can change its optical properties when exposed to laser light. In this way, a shield might be able to stop a laser blast as well as a particle beam or cannon fire. At present, however, photochromatics that can stop laser beams do not exist.

Magnetic Levitation

In science fiction, force fields have another purpose besides deflecting ray-gun blasts, and that is to serve as a platform to defy gravity. In the movie *Back to the Future,* Michael J. Fox rides a "hover board," which resembles a skateboard except that it floats over the street. Such an antigravity device is impossible given the laws of physics as we know them today (as we will see in Chapter 10). But magnetically enhanced hover boards and hover cars could become a reality in the future, giving us the ability to levitate large objects at will. In the future, if "room-temperature superconductors" become a reality, one might be able to levitate objects using the power of magnetic force fields.

If we place two bar magnets next to each other with north poles opposite each other, the two magnets repel each other. (If we rotate the magnet, so that the north pole is close to the other south pole, then the two magnets attract each other.) This same principle, that north poles repel each other, can be used to lift enormous weights off the ground. Already several nations are building advanced magnetic levitation trains (maglev trains) that hover just above the railroad tracks using ordinary magnets. Because they have zero friction, they can attain record-breaking speeds, floating over a cushion of air.

In 1984 the world's first commercial automated maglev system began operation in the United Kingdom, running from Birmingham International Airport to the nearby Birmingham International railway station. Maglev trains have also been built in Germany, Japan, and Korea, although most of them have not been designed for high velocities. The first commercial maglev train operating at high velocities is the initial operating segment (IOS) demonstration line in Shanghai, which travels at a top speed of 268 miles per hour. The Japanese maglev train in Yamanashi prefecture attained a velocity of 361 miles per hour, even faster than the usual wheeled trains.

But these maglev devices are extremely expensive. One way to increase efficiency would be to use superconductors, which lose all electrical resistance when they are cooled down to near absolute zero. Superconductivity was discovered in 1911 by Heike Onnes. If certain substances are cooled to below 20 K above absolute zero, all electrical resistance is lost. Usually when we cool down the temperature of a metal, its resistance decreases gradually. (This is because random vibrations of the atom impede the flow of electrons in a wire. By reducing the temperature, these random motions are reduced, and hence electricity flows with less resistance.) But much to Onnes's surprise, he found that the resistance of certain materials fell abruptly to zero at a critical temperature.

Physicists immediately recognized the importance of this result. Power lines lose a significant amount of energy by transporting electricity across long distances. But if all resistance could be eliminated, electrical power could be transmitted almost for free. In fact, if electricity were made to circulate in a coil of wire, the electricity would circulate for millions of years, without any reduction in energy. Furthermore, magnets of incredible power could be made with little effort from these enormous electric currents. With these magnets, one could lift huge loads with ease.

Despite all these miraculous powers, the problem with superconductivity is that it is very expensive to immerse large magnets in vats of supercooled liquid. Huge refrigeration plants are required to keep liquids supercooled, making superconducting magnets prohibitively expensive.

But one day physicists may be able to create a "room-temperature superconductor," the holy grail of solid-state physicists. The invention of room-temperature superconductors in the laboratory would spark a second industrial revolution. Powerful magnetic fields capable of lifting cars and trains would become so cheap that hover cars might become economically feasible. With room-temperature superconductors, the fantastic flying cars seen in *Back to the Future, Minority Report,* and *Star Wars* might become a reality.

In principle, one might be able to wear a belt made of superconducting magnets that would enable one to effortlessly levitate off the ground. With such a belt, one could fly in the air like Superman. Room-temperature superconductors are so remarkable that they appear in numerous science fiction novels (such as the Ringworld series written by Larry Niven in 1970).

For decades physicists have searched for room-temperature superconductors without successs. It has been a tedious, hit-or-miss process, testing one material after another. But in 1986 a new class of substances called "high-temperature superconductors" was found that became superconductors at about 90 degrees above absolute zero, or 90 K, creating a sensation in the world of physics. The floodgates seemed to open. Month after month, physicists raced one another to break the next world's record for a superconductor. For a brief moment it seemed as if the possibility of room-temperature superconductors would leap off the pages of science fiction novels and into our living rooms. But after a few years of moving at breakneck speed, research in high-temperature superconductors began to slow down.

At present the world's record for a high-temperature superconductor is held by a substance called mercury thallium barium calcium copper oxide, which becomes superconducting at 138 K (–135°C). This relatively high temperature is still a long way from room temperature. But this 138 K record is still important. Nitrogen liquefies at 77 K, and liquid nitrogen costs about as much as ordinary milk. Hence ordinary liquid nitrogen could be used to cool down these high-temperature superconductors rather cheaply. (Of course, room-temperature superconductors would need no cooling whatsoever.)

Embarrassingly enough, at present there is no theory explaining the properties of these high-temperature superconductors. In fact, a Nobel Prize is awaiting the enterprising physicist who can explain how high-temperature superconductors work. (These high-temperature superconductors are made of atoms arranged in distinctive layers. Many physicists theorize that this layering of the ceramic material makes it possible for electrons to flow freely within each layer, creating a superconductor. But precisely how this is done is still a mystery.)

Because of this lack of knowledge, physicists unfortunately resort to a hit-or-miss procedure to search for new high-temperature superconductors. This means that the fabled room-temperature superconductor may be discovered tomorrow, next year, or not at all. No one knows when, or if, such a substance will ever be found.

But if room-temperature superconductors are discovered, a tidal wave of commercial applications could be set off. Magnetic fields that are a million times more powerful than the Earth's magnetic field (which is .5 gauss) might become commonplace.

One common property of superconductivity is called the Meissner effect. If you place a magnet above a superconductor, the magnet will levitate, as if held upward by some invisible force. (The reason for the Meissner effect is that the magnet has the effect of creating a "mirror-image" magnet within the superconductor, so that the original magnet and the mirror-image magnet repel each other. Another way to see this is that magnetic fields cannot penetrate into a superconductor. Instead, magnetic fields are expelled. So if a magnet is held above a superconductor, its lines of force are expelled by the superconductor, and the lines of force then push the magnet upward, causing it to levitate.)

Using the Meissner effect, one can imagine a future in which the highways are made of these special ceramics. Then magnets placed in our belts or our tires could enable us to magically float to our destination, without any friction or energy loss.

The Meissner effect works only on magnetic materials, such as metals. But it is also possible to use superconducting magnets to levitate nonmagnetic materials, called paramagnets and diamagnets. These substances do not have magnetic properties of their own; they acquire their magnetic properties only in the presence of an external magnetic field. Paramagnets are attracted by an external magnet, while diamagnets are repelled by an external magnet.

Water, for example, is a diamagnet. Since all living things are made of water, they can levitate in the presence of a powerful magnetic field. In a magnetic field of about 15 teslas (30,000 times the Earth's field), scientists have levitated small animals, such as frogs. But if room-temperature superconductors become a reality, it should be possible to levitate large nonmagnetic objects as well, via their diamagnetic property.

In conclusion, force fields as commonly described in science fiction do not fit the description of the four forces of the universe. Yet it may be possible to simulate many of the properties of force fields by using a multilayered shield, consisting of plasma windows, laser curtains, carbon nanotubes, and photochromatics. But developing such a shield could be many decades, or even a century, away. And if room-temperature superconductors can be found, one might be able to use powerful magnetic fields to levitate cars and trains and soar in the air, as in science fiction movies.

Given these considerations, I would classify force fields as a Class I impossibility–that is, something that is impossible by today's technology, but possible, in modified form, within a century or so.

2: INVISIBILITY

You cannot depend on your eyes when
your imagination is out of focus.
—MARK TWAIN

In *Star Trek IV: The Voyage Home*, a Klingon battle cruiser is hijacked by the crew of the *Enterprise*. Unlike the starships in the Federation Star Fleet, the starships of the Klingon Empire have a secret "cloaking device" that renders them invisible to light or radar, so that Klingon ships can sneak up behind Federation starships and ambush them with impunity. This cloaking device has given the Klingon Empire a strategic advantage over the Federation of Planets.

Is such a device really possible? Invisibility has long been one of the marvels of science fiction and fantasy, from the pages of *The Invisible Man*, to the magic invisibility cloak of the Harry Potter books, or the ring in *The Lord of the Rings*. Yet for at least a century, physicists have dismissed the possibility of invisibility cloaks, stating flatly that they are impossible: They violate the laws of optics and do not conform to any of the known properties of matter.

But today the impossible may become possible. New advances in "metamaterials" are forcing a major revision of optics textbooks. Working prototypes of such materials have actually been built in the laboratory, sparking intense interest by the media, industry, and the military in making the visible become invisible.

Invisibility Throughout History

Invisibility is perhaps one of the oldest concepts in ancient mythology. Since the advent of recorded history, people who have been alone on a creepy night have been frightened by the invisible spirits of the dead, the souls of the long-departed lurking in the dark. The Greek hero Perseus was able to slay the evil Medusa armed with the helmet of invisibility. Military generals have dreamed of an invisibility cloaking device. Being invisible, one could easily penetrate enemy lines and capture the enemy by surprise. Criminals could use invisibility to pull off spectacular robberies.

Invisibility played a central part in Plato's theory of ethics and morality. In his philosophical masterpiece, *The Republic*, Plato recounts the myth of the ring of Gyges. The poor but honest shepherd Gyges of Lydia enters a hidden cave and finds a tomb containing a corpse wearing a golden ring. Gyges discovers that this golden ring has the magical power to make him invisible. Soon this poor shepherd is intoxicated with the power this ring gives him. After sneaking into the king's palace, Gyges uses his power to seduce the queen and, with her help, murder the king and become the next King of Lydia.

The moral that Plato wished to draw out is that no man can resist the temptation of being able to steal and kill at will. All men are corruptible. Morality is a social construct imposed from the outside. A man may appear to be moral in public to maintain his reputation for integrity and honesty, but once he possesses the power of invisibility, the use of such power would be irresistible. (Some believe that this morality tale was the inspiration for J. R. R. Tolkien's Lord of the Rings trilogy, in which a ring that grants the wearer invisibility is also a source of evil.)

Invisibility is also a common plot device in science fiction. In the *Flash Gordon* series of the 1930s, Flash becomes invisible in order to escape the firing squad of Ming the Merciless. In the Harry Potter novels and movies, Harry dons a special cloak that allows him to roam Hogwarts Castle undetected.

H. G. Wells put much of this mythology into concrete form with his classic novel *The Invisible Man,* in which a medical student accidentally discovers the power of the fourth dimension and becomes invisible. Unfortunately, he uses this fantastic power for private gain, starts a wave of petty crimes, and eventually dies desperately trying to evade the police.

Maxwell's Equations and the Secret of Light

It was not until the work of Scottish physicist James Clerk Maxwell, one of the giants of nineteenth-century physics, that physicists had a firm understanding of the laws of optics. Maxwell, in some sense, was the opposite of Michael Faraday. Whereas Faraday had superb experimental instincts but no formal training whatsoever, Maxwell, a contemporary of Faraday, was a master of advanced mathematics. He excelled as a student of mathematical physics at Cambridge, where Isaac Newton had done his work two centuries earlier.

Newton had invented the calculus, which was expressed in the language of "differential equations," which describe how objects smoothly undergo infinitesimal changes in space and time. The motion of ocean waves, fluids, gases, and cannon balls could all be expressed in the language of differential equations. Maxwell set out with a clear goal, to express the revolutionary findings of Faraday and his force fields through precise differential equations.

Maxwell began with Faraday's discovery that electric fields could turn into magnetic fields and vice versa. He took Faraday's depictions of force fields and rewrote them in the precise language of differential equations, producing one of the most important series of equations in modern science. They are a series of eight fierce-looking differential equations. Every physicist and engineer in the world has to sweat over them when mastering electromagnetism in graduate school.

Next, Maxwell asked himself the fateful question: if magnetic fields can turn into electric fields and vice versa, what happens if they are constantly turning into each other in a never-ending pattern? Maxwell found that these electric-magnetic fields would create a wave, much like an ocean wave. To his astonishment, he calculated the speed of these waves and found it to be the speed of light! In 1864, upon discovering this fact, he wrote prophetically: "This velocity is so nearly that of light that it seems we have strong reason to conclude that light itself . . . is an electromagnetic disturbance."

It was perhaps one of the greatest discoveries in human history. For the first time the secret of light was finally revealed. Maxwell suddenly realized that everything from the brilliance of the sunrise, the blaze of the setting sun, the dazzling colors of the rainbow, and the firmament of stars in the heavens could be described by the waves he was scribbling on a sheet of paper. Today we realize that the entire electromagnetic spectrum–from radar to TV, infrared light, visible light, ultraviolet light, X-rays, microwaves, and gamma rays–is nothing but Maxwell waves, which in turn are vibrating Faraday force fields.

Commenting on the importance of Maxwell's equations, Einstein wrote that they are "the most profound and the most fruitful that physics has experienced since the time of Newton."

(Tragically, Maxwell, one of the greatest physicists of the nineteenth century, died at the early age of forty-eight of stomach cancer, probably the very same disease that killed his mother at the same age. If he had lived longer, he might have discovered that his equations allowed for distortions of space-time that would lead directly to Einstein's relativity theory. It is staggering to realize that relativity might possibly have been discovered at the time of the American Civil War had Maxwell lived longer.)

Maxwell's theory of light and the atomic theory give simple explanations for optics and invisibility. In a solid, the atoms are tightly packed, while in a liquid or gas the molecules are spaced much farther apart. Most solids are opaque because light rays cannot pass through the dense matrix of atoms in a solid, which act like a brick wall. Many liquids and gases, by contrast, are transparent because light can pass more readily between the large spaces between their atoms, a space that is larger than the wavelength of visible light. For example, water, alcohol, ammonia, acetone, hydrogen peroxide, gasoline, and so forth are all transparent, as are gases such as oxygen, hydrogen, nitrogen, carbon dioxide, methane, and so on.

There are some important exceptions to this rule. Many crystals are both solid and transparent. But the atoms of a crystal are arranged in a precise lattice structure, stacked in regular rows, with regular spacing between them. Hence there are many pathways that a light beam may take through a crystalline lattice. Therefore, although a crystal is as tightly packed as any solid, light can still work its way through the crystal.

Under certain circumstances, a solid object may become transparent if the atoms are arranged randomly. This can be done by heating certain materials to a high temperature and then rapidly cooling them. Glass, for example, is a solid with many properties of a liquid because of the random arrangement of its atoms. Certain candies can become transparent via this method as well.

Clearly, invisibility is a property that arises at the atomic level, via Maxwell's equations, and hence would be exceedingly difficult, if not impossible, to duplicate using ordinary means. To make Harry Potter invisible, one would have to liquefy him, boil him to create steam, crystallize him, heat him again, and then cool him, all of which would be quite difficult to accomplish, even for a wizard.

The military, unable to create invisible airplanes, has tried to do the next best thing: create stealth technology, which renders airplanes invisible to radar. Stealth technology relies on Maxwell's equations to create a series of tricks. A stealth fighter jet is perfectly visible to the human eye, but its radar image on an enemy radar screen is only the size of a large bird. (Stealth technology is actually a hodgepodge of tricks. By changing the materials within the jet fighter, reducing its steel content and using plastics and resins instead, changing the angles of its fuselage, rearranging its exhaust pipes, and so on, one can make enemy radar beams hitting the craft disperse in all directions, so they never get back to the enemy radar screen. Even with stealth technology, a jet fighter is not totally invisible; rather, it has deflected and dispersed as much radar as is technically possible.)

Metamaterials and Invisibility

But perhaps the most promising new development involving invisibility is an exotic new material called a "metamaterial," which may one day render objects truly invisible. Ironically, the creation of metamaterials was once thought to be impossible because they violated the laws of optics. But in 2006 researchers at Duke University in Durham, North Carolina, and Imperial College in London successfully defied conventional wisdom and used metamaterials to make an object invisible to microwave radiation. Although there are still many hurdles to overcome, for the first time in history we now have a blueprint to render ordinary objects invisible. (The Pentagon's Defense Advanced Research Projects Agency [DARPA] funded this research.)

Nathan Myhrvold, former chief technology officer at Microsoft, says the revolutionary potential of metamaterials "will completely change the way we approach optics and nearly every aspect of electronics . . . Some of these metamaterials can perform feats that would have seemed miraculous a few decades ago."

What are these metamaterials? They are substances that have optical properties not found in nature. Metamaterials are created by embedding tiny implants within a substance that force electromagnetic waves to bend in unorthodox ways. At Duke University, scientists embedded tiny electrical circuits within copper bands that are arranged in flat, concentric circles (somewhat resembling the coils of an electric oven). The result was a sophisticated mixture of ceramic, Teflon, fiber composites, and metal components. These tiny implants in the copper make it possible to bend and channel the path of microwave radiation in a specific way. Think about the way a river flows around a boulder. Because the water quickly wraps around the boulder, the presence of the boulder has been washed out downstream. Similarly, metamaterials can continuously alter and bend the path of microwaves so that they flow around a cylinder, for example, essentially making everything inside the cylinder invisible to microwaves. If the metamaterial can eliminate all reflection and shadows, then it can render an object totally invisible to that form of radiation.

Scientists successfully demonstrated this principle with a device made of ten fiberglass rings covered with copper elements. A copper ring inside the device was rendered nearly invisible to microwave radiation, casting only a minuscule shadow.

At the heart of metamaterials is their ability to manipulate something called the "index of refraction." Refraction is the bending of light as it moves through transparent media. If you put your hand in water, or look through the lens of your glasses, you notice that water and glass distort and bend the path of ordinary light.

The reason that light bends in glass or water is that light slows down when it enters a dense, transparent medium. The speed of light in a pure vacuum always remains the same, but light traveling through glass or water must pass through trillions of atoms and hence slows down. (The speed of light divided by the slower speed of light inside the medium is called the index of refraction. Since light slows down in glass, the index of refraction is always greater than 1.0). For example, the index of refraction is 1.00 for a vacuum, 1.0003 for air, 1.5 for glass, and 2.4 for diamond. Usually, the denser the medium, the greater the degree of bending, and the greater the index of refraction.

A familiar example of the index of refraction is a mirage. If you are driving on a hot day and look straight toward the horizon, the road may seem to be shimmering, creating the illusion of a glistening lake. In the desert one can sometimes see the outlines of distant cities and mountains on the horizon. This is because hot air rising from the pavement or desert has a lower density than normal air, and hence a lower index of refraction than the surrounding, colder air, and therefore light from distant objects can be refracted off the pavement into your eye, giving you the illusion that you are seeing distant objects.

Usually, the index of refraction is a constant. A narrow beam of light is bent when it enters glass and then keeps going in a straight line. But assume for the moment that you could control the index of refraction at will, so that it could change continuously at every point in the glass. As light moved in this new material, light could bend and meander in new directions, creating a path that would wander throughout the substance like a snake.

If one could control the index of refraction inside a metamaterial so that light passed around an object, then the object would become invisible. To do this, this metamaterial must have a *negative* index of refraction, which every optics textbook says is impossible. (Metamaterials were first theorized in a paper by Soviet physicist Victor Veselago in 1967 and were shown to have weird optical properties, such as a negative refractive index and reversed Doppler effect. Metamaterials are so bizarre and preposterous that they were once thought to be impossible to construct. But in the last few years, metamaterials have actually been manufactured in the laboratory, forcing reluctant physicists to rewrite all the textbooks on optics.)

Researchers in metamaterials are constantly pestered by journalists who wish to know when invisibility cloaks will hit the market. The answer is: not anytime soon.

David Smith of Duke University says, "Reporters, they call up and they just want you to say a number. Number of months, number of years. They push and push and push and you finally say, well, maybe fifteen years. Then you've got your headline, right? Fifteen years till Harry Potter's cloak." That's why he now declines to give any specific timetable. Fans of *Harry Potter* or *Star Trek* may have to wait. While a true invisibility cloak *is* possible within the laws of physics, as most physicists will agree, formidable technical hurdles remain before this technology can be extended to work with visible light rather than just microwave radiation.

In general, the internal structures implanted inside the metamaterial must be smaller than the wavelength of the radiation. For example, microwaves can have a wavelength of about 3 centimeters, so for a metamaterial to bend the path of microwaves, it must have tiny implants embedded inside it that are smaller than 3 centimeters. But to make an object invisible to green light, with a wavelength of 500 nanometers (nm), the metamaterial must have structures embedded within it that are only about 50 nanometers long–and nanometers are atomic-length scales requiring nanotechnology. (One nanometer is a billionth of a meter in length. Approximately five atoms can fit within a single nanometer.) This is perhaps the key problem we face in our attempts to create a true invisibility cloak. The individual atoms inside a metamaterial would have to be modified to bend a light beam like a snake.

Metamaterials for Visible Light

The race is on.

Ever since the announcement that metamaterials have been fabricated in the laboratory there has been a stampede of activity in this area, with new insights and startling breakthroughs coming every few months. The goal is clear: to use nanotechnology to create metamaterials that can bend visible light, not just microwaves. Several approaches have been proposed, all of them quite promising.

One proposal is to use off-the-shelf technology, that is, to borrow known techniques from the semiconductor industry to create new metamaterials. A technique called "photolithography" lies at the heart of computer miniaturization and hence drives the computer revolution. This technology enables engineers to place hundreds of millions of tiny transistors onto a silicon wafer no bigger than your thumb.

The reason that computer power doubles every eighteen months (which is called Moore's law) is because scientists use ultraviolet radiation to "etch" tinier and tinier components onto a silicon chip. This technique is very similar to the way in which stencils are used to create colorful T-shirts. (Computer engineers start with a thin wafer and then apply extremely thin coatings of various materials on top. A plastic mask is then placed over the wafer, which acts as a template. It contains the complex outlines of the wires, transistors, and computer components that are the basic skeleton of the circuitry. The wafer is then bathed in ultraviolet radiation, which has a very short wavelength, and that radiation imprints the pattern onto the photosensitive wafer. By treating the wafer with special gases and acids, the complex circuitry of the mask is etched onto the wafer where it was exposed to ultraviolet light. This process creates a wafer containing hundreds of millions of tiny grooves, which form the outlines of the transistors.) At present, the smallest components that one can create with this etching process are about 30 nm (or about 150 atoms across).

A milestone in the quest for invisibility came when this silicon wafer etching technology was used by a group of scientists to create the first metamaterial that operates in the visible range of light. Scientists in Germany and at the U.S. Department of Energy announced in early 2007 that, for the first time in history, they had fabricated a metamaterial that worked for red light. The "impossible" had been achieved in a remarkably short time.

Physicist Costas Soukoulis of the Ames Laboratory in Iowa, with Stefan Linden, Martin Wegener, and Gunnar Dolling of the University of Karlsruhe, Germany, were able to create a metamaterial that had an index of –.6 for red light, at a wavelength of 780 nm. (Previously, the world record for radiation bent by a metamaterial was 1,400 nm, which put it outside the range of visible light, in the range of infrared.)

The scientists first started with a glass sheet, and then deposited a thin coating of silver, magnesium fluoride, and then another layer of silver, forming a "sandwich" of fluoride that was only 100 nm thick. Then, using standard etching techniques, they created a large array of microscopic square holes in the sandwich, creating a grid pattern resembling a fishnet. (The holes are only 100 nm wide, much smaller than the wavelength of red light.) Then they passed a red light beam through the material and measured its index, which was –.6.

These physicists foresee many applications of this technology. Metamaterials "may one day lead to the development of a type of flat superlens that operates in the visible spectrum," says Dr. Soukoulis. "Such a lens would offer superior resolution over conventional technology, capturing details much smaller than one wavelength of light." The immediate application of such a "superlens" would be to photograph microscopic objects with unparalleled clarity, such as the inside of a living human cell, or to diagnose diseases in a baby inside the womb. Ideally one would be able to obtain photographs of the components of a DNA molecule without having to use clumsy X-ray crystallography.

So far these scientists have demonstrated a negative index of refraction only for red light. Their next step would be to use this technology to create a metamaterial that would bend red light entirely around an object, rendering it invisible to that light.

Future developments along these lines may occur in the area of "photonic crystals." The goal of photonic crystal technology is to create a chip that uses light, rather than electricity, to process information. This entails using nanotechnology to etch tiny components onto a wafer, such that the index of refraction changes with each component. Transistors using light have several advantages over those using electricity. For example, there is much less heat loss for photonic crystals. (In advanced silicon chips, the heat generated is enough to fry an egg. Thus they must be continually cooled down or else they will fail, and keeping them cool is very costly.) Not surprisingly, the science of photonic crystals is ideally suited for metamaterials, since both technologies involve manipulating the index of refraction of light at the nanoscale.

Invisibility via Plasmonics

Not to be outdone, yet another group announced in mid-2007 that they have created a metamaterial that bends visible light using an entirely different technology, called "plasmonics." Physicists Henri Lezec, Jennifer Dionne, and Harry Atwater at the California Institute of Technology announced that they had created a metamaterial that had a negative index for the more difficult blue-green region of the visible spectrum of light.

The goal of plasmonics is to "squeeze" light so that one can manipulate objects at the nanoscale, especially on the surface of metals. The reason metals conduct electricity is that electrons are loosely bound to metal atoms, so they can freely move along the surface of the metal lattice. The electricity flowing in the wires in your home represents the smooth flow of these loosely bound electrons on the metal surface. But under certain conditions, when a light beam collides with the metal surface, the electrons can vibrate in unison with the original light beam, creating wavelike motions of the electrons on the metal surface (called plasmons), and these wavelike motions beat in unison with the original light beam. More important, one can "squeeze" these plasmons so that they have the same frequency as the original beam (and hence carry the same information) but have a much smaller wavelength. In principle, one might then cram these squeezed waves onto nanowires. As with photonic crystals, the ultimate goal of plasmonics is to create computer chips that compute using light, rather than electricity.

The Cal Tech group built their metamaterial out of two layers of silver, with a silicon-nitrogen insulator in between (with a thickness of only 50 nm), which acted as a "waveguide" that could shepherd the direction of the plasmonic waves. Laser light enters and exits the apparatus via two slits carved into the metamaterial. By analyzing the angles at which the laser light is bent as it passes through the metamaterial, one can then verify that the light is being bent via a negative index.

The Future of Metamaterials

Progress in metamaterials will accelerate in the future for the simple reason that there is already intense interest in creating transistors that use light beams rather than electricity. Research in invisibility can therefore "piggyback" on the ongoing research in photonic crystals and plasmonics for creating replacements for the silicon chip. Already hundreds of millions of dollars are being invested in creating replacements for silicon technology, and research in metamaterials will benefit from these research efforts.

With breakthroughs occurring in this field every few months, it's not surprising that some physicists see some sort of practical invisibility shield emerging out of the laboratory perhaps within a few decades. In the next few years, for example, scientists are confident that they will be able to create metamaterials that can render an object totally invisible for one frequency of visible light, at least in two dimensions. To do this would require embedding tiny nano implants not in regular arrays, but in sophisticated patterns so that light would bend smoothly around an object.

Next, scientists will have to create metamaterials that can bend light in three dimensions, not just for flat two-dimensional surfaces. Photolithography has been perfected for making flat silicon wafers, but creating three-dimensional metamaterials will require stacking wafers in a complex fashion.

After that, scientists will have to solve the problem of creating metamaterials that can bend not just one frequency but many. This will be perhaps the most difficult task, since the tiny implants that have been devised so far bend light of only one precise frequency. Scientists may have to create metamaterials based on layers, with each layer bending a specific frequency. The solution to this problem is not clear.

Nevertheless, once an invisibility shield is finally made, it might be a clunky device. Harry Potter's cloak was made of thin, flexible cloth and rendered anyone draped inside invisible. But for this to be possible the index of refraction inside the cloth would have to be constantly changing in complex ways as it fluttered, which is impractical. More than likely a true invisibility "cloak" would have to be made of a solid cylinder of metamaterials, at least initially. That way the index of refraction could be fixed inside the cylinder. (More advanced versions could eventually incorporate metamaterials that are flexible and can twist and still make light flow within the metamaterials on the correct path. In this way, anyone inside the cloak would have some flexibility of movement.)

Some have pointed out a flaw in the invisibility shield: anyone inside would not be able to look outside without becoming visible. Imagine Harry Potter being totally invisible except for his eyes, which appear to be floating in midair. Any eye holes on the invisibility cloak would be clearly visible from the outside. If Harry Potter were totally invisible, then he would be sitting blindly beneath his invisibility cloak. (One possible solution to this problem might be to insert two tiny glass plates near the location of the eye holes. These glass plates would act as "beam splitters," splitting off a tiny portion of the light hitting the plates, and then sending the light into the eyes. So most of the light hitting the cloak would flow around it, rendering the person invisible, but a tiny amount of light would be diverted into the eyes.)

As daunting as these difficulties are, scientists and engineers are optimistic that an invisibility shield of some sort can be built in the coming decades.

Invisibility and Nanotechnology

As I mentioned earlier, the key to invisibility may be nanotechnology, that is, the ability to manipulate atomic-sized structures about a billionth of a meter across.

The birth of nanotechnology dates back to a famous 1959 lecture given by Nobel laureate Richard Feynman to the American Physical Society, with the tongue-in-cheek title "There's Plenty of Room at the Bottom." In that lecture he speculated on what the smallest machines might look like, consistent with the known laws of physics. He realized that machines could be built smaller and smaller until they hit atomic distances, and then atoms could be used to create other machines. Atomic machines, such as pulleys, levers, and wheels, were well within the laws of physics, he concluded, though they would be exceedingly difficult to make.

Nanotechnology languished for years, because manipulating individual atoms was beyond the technology of the time. But then physicists made a breakthrough in 1981, with the invention of the scanning tunneling microscope, which won the Nobel Prize in Physics for scientists Gerd Binnig and Heinrich Rohrer, working at the IBM lab in Zurich.

Suddenly physicists were able to obtain stunning "pictures" of individual atoms arrayed just as in the chemistry books, something that critics of the atomic theory once considered impossible. Gorgeous photographs of atoms lined up in a crystal or metal were now possible. The chemical formulae used by scientists, with a complex series of atoms wrapped up in a molecule, could be seen with the naked eye. Moreover, the scanning tunneling microscope made possible the manipulation of individual atoms. In fact, the letters "IBM" were spelled out via individual atoms, creating quite a stir in the scientific world. Scientists were no longer blind when manipulating individual atoms, but could actually see and play with them.

The scanning tunneling microscope is deceptively simple. Like a phonograph needle scanning a disk, a sharp probe is passed slowly over the material to be analyzed. (The tip is so sharp that it consists of only a single atom.) A small electrical charge is placed on the probe, and a current flows from the probe, through the material, and to the surface below. As the probe passes over an individual atom, the amount of current flowing through the probe varies, and the variations are recorded. The current rises and falls as the needle passes over an atom, thereby tracing its outline in remarkable detail. After many passes, by plotting the fluctuations in the current flows, one is able to obtain beautiful pictures of the individual atoms making up a lattice.

(The scanning tunneling microscope is made possible by a strange law of quantum physics. Normally electrons do not have enough energy to pass from the probe, through the substance, to the underlying surface. But because of the uncertainty principle, there is a small probability that the electrons in the current will "tunnel" or penetrate through the barrier, even though this is forbidden by Newtonian theory. Thus the current that flows through the probe is sensitive to tiny quantum effects in the material. I will discuss the effects of the quantum theory later in more detail.)

The probe is also sensitive enough to move individual atoms around, to create simple "machines" out of individual atoms. The technology is so advanced now that a cluster of atoms can be displayed on a computer screen and then, by simply moving the cursor of the computer, the atoms can be moved around any way you want. You can manipulate scores of atoms at will as if playing with Lego blocks. Besides spelling out the letters of the alphabet using individual atoms, one can also create atomic toys, such as an abacus made out of individual atoms. The atoms are arrayed on a surface, with vertical slots. Inside these vertical slots one can insert carbon Buckyballs (shaped like a soccer ball, but made of individual carbon atoms). These carbon balls can then be moved up and down each slot, thereby making an atomic abacus.

It is also possible to carve atomic devices using electron beams. For example, scientists at Cornell University have made the world's smallest guitar, one that is twenty times smaller than a human hair, carved out of crystalline silicon. It has six strings, each one hundred atoms thick, and the strings can be plucked using an atomic force microscope. (This guitar will actually play music, but the frequencies it produces are well above the range of the human ear.)

At present, most of these nanotech "machines" are mere toys. More complicated machines with gears and ball bearings have yet to be created. But many engineers feel confident that the time is coming when we will be able to produce true atomic machines. (Atomic machines are actually found in nature. Cells can swim freely in water because they can wiggle tiny hairs. But when one analyzes the joint between the hair and the cell, one sees that it is actually an atomic machine that allows the hair to move in all directions. So one key to developing nanotechnology is to copy nature, which mastered the art of atomic machines billions of years ago.)

Holograms and Invisibility

Another way to render a person partially invisible is to photograph the scenery behind a person and then project that background image directly onto the person's clothes or onto a screen in front of him. As seen from the front, it appears as if the person has become transparent, that light has somehow passed right through the person's body.

Naoki Kawakami, of the Tachi Laboratory at the University of Tokyo, has been hard at work on this process, which is called "optical camouflage." He says, "It would be used to help pilots see through the floor of the cockpit at a runway below, or for drivers trying to see through a fender to park a car." Kawakami's "cloak" is covered with tiny light-reflective beads that act like a movie screen. A video camera photographs what is behind the cloak. Then this image is fed into a video projector that lights up the front of the cloak, so it appears as if light has passed through the person.

Prototypes of the optical camouflage cloak actually exist in the lab. If you look directly at a person wearing this screenlike cloak, it appears as if the person has disappeared, because all you see is the image behind the person. But if you move your eyes a bit, the image on the cloak does not change, which tells you that it is a fake. A more realistic optical camouflage would need to create the illusion of a 3-D image. For this, one would need holograms.

A hologram is a 3-D image created by lasers (like the 3-D image of Princess Leia in *Star Wars*). A person could be rendered invisible if the background scenery was photographed with a special holographic camera and the holographic image was then projected out through a special holographic screen placed in front of the person. A viewer standing in front of that person would see the holographic screen, containing the 3-D image of the background scenery, minus the person. It would appear as if the person had disappeared. In that person's place would be a precise 3-D image of the background scenery. Even if you moved your eyes, you would not be able to tell that what you were seeing was fake.

These 3-D images are made possible because laser light is "coherent," that is, all the waves are vibrating in perfect unison. Holograms are produced by making a coherent laser beam split in two pieces. Half of the beam shines on a photographic film. The other half illuminates an object, bounces off, and then shines on the same photographic film. When these two beams interfere on the film, an interference pattern is created that encodes all the information of the original 3-D wave. The film, when developed, doesn't look like much, just an intricate spiderweb pattern of whirls and lines. But when a laser beam is allowed to shine on this film, an exact 3-D replica of the original object suddenly appears as if by magic.

The technical problems with holographic invisibility are formidable, however. One challenge is to create a holographic camera that is capable of taking at least 30 frames per second. Another problem is storing and processing all the information. Finally, one would need to project this image onto a screen so that the image looks realistic.

Invisibility via the Fourth Dimension

We should also mention that an even more sophisticated way of becoming invisible was mentioned by H. G. Wells in *The Invisible Man,* and it involved using the power of the fourth dimension. (Later in the book I will discuss in more detail the possible existence of higher dimensions.) Could we perhaps leave our three-dimensional universe and hover over it from the vantage point of a fourth dimension? Like a three-dimensional butterfly hovering over a two-dimensional sheet of paper, we would be invisible to anyone living in the universe below us. One problem with this idea is that higher dimensions have not yet been proven to exist. Moreover, a hypothetical journey to a higher dimension would require energies far beyond anything attainable with our current technology. As a viable way to achieve invisibility, this method is clearly beyond our knowledge and ability today.

Given the enormous strides made so far in achieving invisibility, it clearly qualifies as a Class I impossibility. Within the next few decades, or at least within this century, a form of invisibility may become commonplace.

3: PHASERS AND DEATH STARS

Radio has no future. Heavier-than-air flying machines are impossible. X-rays will prove to be a hoax.

—PHYSICIST LORD KELVIN, 1899

The (atomic) bomb will never go off.
I speak as an expert in explosives.

—ADMIRAL WILLIAM LEAHY

4–3–2–1, fire!

The Death Star is a colossal weapon, the size of an entire moon. Firing point-blank at the helpless planet Alderaan, home world of Princess Leia, the Death Star incinerates it, causing it to erupt in a titanic explosion, sending planetary debris hurtling throughout the solar system. A billion souls scream out in anguish, creating a disturbance in the Force felt throughout the galaxy.

But is the Death Star weapon of the *Star Wars* saga really possible? Could such a weapon channel a battery of laser cannons to vaporize an entire planet? What about the famous light sabers wielded by Luke Skywalker and Darth Vader that can slice through reinforced steel yet are made of beams of light? Are ray guns, like the phasers in *Star Trek*, viable weapons for future generations of law enforcement officers and soldiers?

In *Star Wars* millions of moviegoers were dazzled by these original, stunning special effects, but they fell flat for some critics, who panned them, stating that all this was in good fun, but it was patently impossible. Moon-sized, planet-busting ray guns are outlandish, and so are swords made of solidified light beams, even for a galaxy far, far away, they chanted. George Lucas, the master of special effects, must have gotten carried away this time.

Although this may be difficult to believe, the fact is there is no physical limit to the amount of raw energy that can be crammed onto a light beam. There is no law of physics preventing the creation of a Death Star or light sabers. In fact, planet-busting beams of gamma radiation exist in nature. The titanic burst of radiation from a distant gamma ray burster in deep space creates an explosion second only to the big bang itself. Any planet unfortunate enough to be within the crosshairs of a gamma ray burster will indeed be fried or blown to bits.

Beam Weapons Through History

The dream of harnessing beams of energy is actually not new but is rooted in ancient mythology and lore. The Greek god Zeus was famous for unleashing lightning bolts on mortals. The Norse god Thor had a magic hammer, Mjolnir, which could fire bolts of lightning, while the Hindu god Indra was known for firing beams of energy from a magic spear.

The concept of using rays as a practical weapon probably began with the work of the great Greek mathematician Archimedes, perhaps the greatest scientist in all of antiquity, who discovered a crude version of calculus two thousand years ago, before Newton and Leibniz. In one legendary battle against the forces of Roman general Marcellus during the Second Punic War in 214 BC, Archimedes helped to defend the kingdom of Syracuse and is believed to have created large batteries of solar reflectors that focused the sun's rays onto the sails of enemy ships, setting them ablaze. (There is still debate even today among scientists as to whether this was a practical, working beam weapon; various teams of scientists have tried to duplicate this feat with differing results.)

Ray guns burst onto the science fiction scene in 1889 with H. G. Wells's classic *War of the Worlds,* in which aliens from Mars devastate entire cities by shooting beams of heat energy from weapons mounted on their tripods. During World War II, the Nazis, always eager to exploit the latest advances in technology to conquer the world, experimented with various forms of ray guns, including a sonic device, based on parabolic mirrors, that could focus intense beams of sound.

Weapons created from focused light beams entered the public imagination with the James Bond movie *Goldfinger,* the first Hollywood film to feature a laser. (The legendary British spy was strapped onto a metal table as a powerful laser beam slowly advanced, gradually melting the table between his legs and threatening to slice him in half.)

Physicists originally scoffed at the idea of the ray guns featured in Wells's novel because they violated the laws of optics. According to Maxwell's equations, the light we see around us rapidly disperses and is incoherent (i.e., it is a jumble of waves of different frequencies and phases). It was once thought that coherent, focused, uniform beams of light, as we find with laser beams, were impossible to create.

The Quantum Revolution

All this changed with the coming of the quantum theory. At the turn of the twentieth century it was clear that although Newton's laws and Maxwell's equations were spectacularly successful in explaining the motion of the planets and the behavior of light, they could not explain a whole class of phenomena. They failed miserably to explain why materials conduct electricity, why metals melt at certain temperatures, why gases emit light when heated, why certain substances become superconductors at low temperatures–all of which requires an understanding of the internal dynamics of atoms. The time was ripe for a revolution. Two hundred and fifty years of Newtonian physics was about to be overthrown, heralding the birth pangs of a new physics.

In 1900 Max Planck in Germany proposed that energy was not continuous, as Newton thought, but occurred in small, discrete packets, called "quanta." Then in 1905 Einstein postulated that light consisted of these tiny discrete packets (or quanta), later dubbed "photons." With this powerful but simple idea Einstein was able to explain the photoelectric effect, why electrons are emitted from metals when you shine a light on them. Today the photoelectric effect and the photon form the basis of TV, lasers, solar cells, and much of modern electronics. (Einstein's theory of the photon was so revolutionary that even Max Planck, normally a great supporter of Einstein, could not at first believe it. Writing about Einstein, Planck said, "That he may sometimes have missed the target . . . as for example, in his hypothesis of light quanta, cannot really be held against him.")

Then in 1913 the Danish physicist Niels Bohr gave us an entirely new picture of the atom, one that resembled a miniature solar system. But unlike in a solar system in outer space, electrons can only move in discrete orbits or shells around the nucleus. When electrons "jumped" from one shell to a smaller shell with less energy, they emitted a photon of energy. When an electron absorbed a photon of a discrete energy, it "jumped" to a larger shell with more energy.

A nearly complete theory of the atom emerged in 1925, with the coming of quantum mechanics and the revolutionary work of Erwin Schrödinger, Werner Heisenberg, and many others. According to the quantum theory, the electron was a particle, but it had a wave associated with it, giving it both particle- and wavelike properties. The wave obeyed an equation, called the Schrödinger wave equation, which enabled one to calculate the properties of atoms, including all the "jumps" postulated by Bohr.

Before 1925 atoms were still considered mysterious objects that many, like philosopher Ernst Mach, believed might not exist at all. After 1925 one could actually peer deep into the dynamics of the atom and actually predict its properties. Astonishingly, this meant that if you had a big enough computer, you could derive the properties of the chemical elements from the laws of the quantum theory. In the same way that Newtonian physicists could compute the motions of all the celestial bodies in the universe if they had a big enough calculating machine, quantum physicists claimed that they could in principle compute all the properties of the chemical elements of the universe. If one had a big enough computer, one could also write the wave function of an entire human being.

Masers and Lasers

In 1953 Professor Charles Townes of the University of California at Berkeley and his colleagues produced the first coherent radiation in the form of microwaves. It was christened the "maser" (for microwave amplification through stimulated emission of radiation). He and Russian physicists Nikolai Basov and Aleksandr Prokhorov would eventually win the Nobel Prize in 1964. Soon their results were extended to visible light, giving birth to the laser. (A phaser, however, is a fictional device popularized in *Star Trek*.)

In a laser you first begin with a special medium that will transmit the laser beam, such as a special gas, crystal, or diode. Then you pump energy into this medium from the outside, in the form of electricity, radio, light, or a chemical reaction. This sudden influx of energy pumps up the atoms of the medium, so the electrons absorb the energy and then jump into the outer electron shells.

In this excited, pumped-up state, the medium is unstable. If one then sends in a light beam through the medium, the photons will hit each atom, causing it to suddenly decay down to a lower level, releasing more photons in the process. This in turn triggers even more electrons to release photons, eventually creating a cascade of collapsing atoms, with trillions upon trillions of photons suddenly released into the beam. The key is that for certain substances, when this avalanche of photons is occurring all the photons are vibrating in unison, that is, they are coherent.

(Picture a line of dominoes. Dominoes in their lowest energy state lie flat on a table. Dominoes in a high-energy, pumped-up state stand up vertically, similar to the pumped-up atoms in the medium. If you push one domino, you can trigger a sudden collapse of all this energy at once, just as in a laser beam.)

Only certain materials will "lase," that is, it is only in special materials that when a photon hits a pumped-up atom a photon will be emitted that is coherent with the original photon. As a result of this coherence, in this flood of photons all the photons are vibrating in unison, creating a pencil-thin laser beam. (Contrary to myth, the laser beam does not stay pencil-thin forever. A laser beam fired onto the moon, for example, will gradually expand until it creates a spot a few miles across.)

A simple gas laser consists of a tube of helium and neon gas. When electricity is sent through the tube the atoms are energized. Then, if the energy is suddenly released all at once, a beam of coherent light is produced. The beam is amplified by using two mirrors, one placed at either end, so the beam bounces back and forth between them. One mirror is completely opaque, but the other allows a tiny amount of light to escape on each pass, producing a beam that shoots out one end.

Today lasers are found almost everywhere, from grocery store checkout stands, to fiber-optic cables carrying the Internet, to laser printers and CD players, to modern computers. They are also used in eye surgery, to remove tattoos, and even in cosmetic salons. Over $5.4 billion worth of lasers were sold worldwide in 2004.

Types of Lasers and Fusion

New lasers are being discovered almost every day as new materials are found that can lase, and as new ways are discovered for pumping energy into the medium.

The question is, are any of these technologies suitable for building a ray gun or a light saber? Is it possible to build a laser powerful enough to energize a Death Star? Today a bewildering variety of lasers exist, depending on the material that lases and the energy that is injected into the material (e.g., electricity, intense beams of light, even chemical explosions). Among them are

- *Gas lasers.* These lasers include helium-neon lasers, which are very common, creating a familiar red beam. They are energized by radio waves or electricity. Helium-neon lasers are quite weak. But carbon dioxide gas lasers can be used for blasting, cutting, and welding in heavy industry and can create beams of enormous power that are totally invisible.
- *Chemical lasers.* These powerful lasers are energized by a chemical reaction, such as a burning jet of ethylene and nitrogen trifluoride, or NF_3. Such lasers are powerful enough to be used in military applications. Chemical lasers are used in the U.S. military's airborne and ground lasers, which can produce millions of watts of power, and are designed to shoot down short-range missiles in midflight.
- *Excimer lasers.* These lasers are also powered by chemical reactions, often involving an inert gas (e.g., argon, krypton, or xenon) and fluorine or chlorine. They produce ultraviolet light and can be used to etch tiny transistors onto chips in the semiconductor industry, or for delicate Lasik eye surgery.
- *Solid-state lasers.* The first working laser ever made consisted of a chromium-sapphire ruby crystal. A large variety of crystals will support a laser beam, in conjunction with yttrium, holmium, thulium, and other chemicals. They can produce high-energy ultrashort pulses of laser light.
- *Semiconductor lasers.* Diodes, which are commonly used in the semiconductor industry, can produce the intense beams used in industrial cutting and welding. They are also often found in checkout stands in grocery stores, reading the bar codes of your grocery items.
- *Dye lasers.* These lasers use organic dyes as their me-

dium. They are exceptionally useful in terms of creating ultrashort pulses of light, often lasting only trillionths of a second.

Lasers and Ray Guns?

Given the enormous variety of commercial lasers and the power of military lasers, why don't we have ray guns available for use in combat and on the battlefield? Ray guns of one sort or another seem to be standard-issue weaponry in science fiction movies. Why aren't we working to create them?

The simple answer is the lack of a portable power pack. One would need miniature power packs that contain the power of a huge electrical power station yet are small enough to fit on your palm. At present the only way to harness the power of a large commercial power station is to build one. At present the smallest portable military device that can contain vast amounts of energy is a miniature hydrogen bomb, which might destroy you as well as the target.

There is a second, ancillary problem as well–the stability of the lasing material. Theoretically, there is no limit to the energy one can concentrate on a laser. The problem is that the lasing material in a handheld ray gun would not be stable. Crystal lasers, for example, will overheat and crack if too much energy is pumped into them. Hence to create an extremely powerful laser, the kind that might vaporize an object or neutralize a foe, one might need to use the power of an explosion. In that case, the stability of the lasing material is not such a limitation, since such a laser would be used only once.

Because of the problems in creating a portable power pack and a stable lasing material, building a handheld ray gun is not possible with today's technology. Ray guns are possible, but only if they are connected by a cable to a power supply. Or perhaps with nanotechnology we might be able to create miniature batteries that store or generate enough energy to create the intense bursts of energy required of a handheld device. At present, as we have seen, nanotechnology is quite primitive. At the atomic level, scientists have been able to create atomic devices that are quite ingenious, but impractical, such as an atomic abacus and an atomic guitar. But it is conceivable that late in this century or the next, nanotechnology may be able to give us miniature batteries that can store such fabulous amounts of energy.

Light sabers suffer from a similar problem. When the movie *Star Wars* first came out in the 1970s and light sabers became a best-selling toy among children, many critics pointed out that such a device could never be made. First, it is impossible to solidify light. Light always travels at the speed of light; it cannot be made solid. Second, light beams do not terminate in midair as do the light sabers used in *Star Wars.* Light beams keep on going forever; a real light saber would stretch into the sky.

Actually there is a way to construct a kind of light saber using plasmas, or superhot ionized gas. Plasmas can be made hot enough to glow in the dark and also slice through steel. A plasma light saber would consist of a thin, hollow rod that slides out of the handle, like a telescope. Inside this tube hot plasmas would be released that would then escape through small holes placed regularly along the rod. As the plasma flowed out of the handle, up the rod, and through the holes, it would create a long, glowing tube of superhot gas, sufficient to melt steel. This device is sometimes referred to as a plasma torch.

So it is possible to create a high-energy device that resembles a light saber. But as with ray guns, you would have to create a high-energy portable power pack. Either you would need long cables connecting the light saber to a power supply, or you would have to create, via nanotechnology, a tiny power supply that could deliver huge amounts of power.

So while ray guns and light sabers are possible to create in some form today, the handheld weapons found in science fiction movies are beyond current technology. But late in this century or the next, with new advances in material science and also nanotechnology, a form of ray gun might be developed, making it a Class I impossibility.

ENERGY FOR A DEATH STAR

To create a Death Star laser cannon that can destroy an entire planet and terrorize a galaxy, such as that described in *Star Wars,* one would need to create the most powerful laser ever conceived. At present some of the most powerful lasers on Earth are being used to unleash temperatures found only in the center of stars. In the form of fusion reactors, they might one day harness the power of the stars on Earth.

Fusion machines try to mimic what happens in outer space when a star first forms. A star begins as a huge ball of formless hydrogen gas, until gravity compresses the gas and thereby heats it up; temperatures eventually reach astronomical levels. Deep inside a star's core, for example, temperatures can soar to between 50 million and 100 million degrees centigrade, hot enough to cause hydrogen nuclei to slam into each other, creating helium nuclei and a burst of energy. The fusion of hydrogen into helium, whereby a small amount of mass is converted into the explosive energy of a star via Einstein's famous equation $E = mc^2$, is the energy source of the stars.

There are two ways in which scientists are currently attempting to harness fusion on the Earth. Both have proven to be much more difficult to develop than expected.

INERTIAL CONFINEMENT FOR FUSION

The first method is called "inertial confinement." It uses the most powerful lasers on Earth to create a piece of the sun in the laboratory. A neodymium glass solid-state laser is ideally suited to duplicate the blistering temperatures found only in the core of a star. These laser systems are the size of a large factory and contain a battery of lasers that shoot a series of parallel laser beams down a long tunnel. These high-power laser beams then strike a series of small mirrors arranged around a sphere; the mirrors carefully focus the laser beams uniformly onto a tiny, hydrogen-rich pellet (made of substances such as lithium deuteride, the active ingredient of a hydrogen bomb). The pellet is usually the size of a pinhead and weighs only 10 milligrams.

The blast of laser light incinerates the surface of the pellet, causing the surface to vaporize and compress the pellet. As the pellet collapses, a shock wave is created that reaches the core of the pellet, sending temperatures soaring to millions of degrees, sufficient to fuse hydrogen nuclei into helium. The temperatures and pressures are so astronomical that "Lawson's criterion" is satisfied, the same criterion that is satisfied in hydrogen bombs and in the core of stars. (Lawson's criterion states that a specific range of temperatures, density, and time of confinement must be attained in order to unleash the fusion process in a hydrogen bomb, in a star, or in a fusion machine.)

In the inertial confinement process vast amounts of energy are released, including neutrons. (The lithium deuteride can hit temperatures of 100 million degrees centigrade and a density twenty times that of lead.) A burst of neutrons is then emitted from the pellet, and the neutrons strike a spherical blanket of material surrounding the chamber, and the blanket is heated up. The heated blanket then boils water, and the steam can be used to power a turbine and produce electricity.

The problem, however, lies in being able to focus such intense power evenly onto a tiny spherical pellet. The first serious attempt at creating laser fusion was the Shiva laser, a twenty-beam laser system built at the Lawrence Livermore National Laboratory (LLNL) in California that began operation in 1978. (Shiva is the Hindu goddess with multiple arms, which the laser system design mimics.) The performance of the Shiva laser system was disappointing, but it was sufficient to prove that laser fusion can technically work. The Shiva laser system was later replaced by the Nova laser, with ten times the energy of Shiva. But the Nova laser also failed to achieve proper ignition of the pellets. Nonetheless, it paved the way for the current research in the National Ignition Facility (NIF), which began construction in 1997 at the LLNL.

The NIF, which is supposed to be operational in 2009, is a monstrous machine, consisting of a battery of 192 laser beams, packing an enormous output of 700 trillion watts of power (the output of about

700,000 large nuclear power plants concentrated in a single burst of energy). It is a state-of-the-art laser system designed to achieve full ignition of the hydrogen-rich pellets. (Critics have also pointed out its obvious military use, since it can simulate the detonation of a hydrogen bomb and perhaps make possible the creation of a new nuclear weapon, the pure fusion bomb, which does not require a uranium or plutonium atomic bomb to kick-start the fusion process.)

But even the NIF laser fusion machine, containing the most powerful lasers on Earth, cannot begin to approximate the devastating power of the *Star Wars* Death Star. To build such a device we must look to other sources of power.

Magnetic Confinement for Fusion

The second method scientists could potentially use to energize a Death Star is called "magnetic confinement," a process in which a hot plasma of hydrogen gas is contained within a magnetic field. In fact, this method could actually provide the prototype for the first commercial fusion reactors. Currently the most advanced fusion project of this type is the International Thermonuclear Experimental Reactor (ITER). In 2006 a coalition of nations (including the European Union, the United States, China, Japan, Korea, Russia, and India) decided to build the ITER in Cadarache, in southern France. It is designed to heat hydrogen gas to 100 million degrees centigrade. It could become the first fusion reactor in history to generate more energy than it consumes. It is designed to generate 500 megawatts of power for 500 seconds (the current record is 16 megawatts of power for 1 second). The ITER should generate its first plasma by 2016 and be fully operational in 2022. At a cost of $12 billion, it is the third most expensive scientific project in history (after the Manhattan Project and the International Space Station).

The ITER looks like a large doughnut, with hydrogen gas circulating inside and huge coils of wire winding around the surface. The coils are cooled down until they become superconducting, and then a huge amount of electrical energy is pumped into them, creating a magnetic field that confines the plasma inside the doughnut. When an electrical current is fed inside the doughnut, the gas is heated to stellar temperatures.

The reason scientists are so excited by the ITER is the prospect of creating a cheap energy source. The fuel supply for fusion reactors is ordinary seawater, which is rich in hydrogen. At least on paper, fusion may provide us with an inexhaustible, cheap supply of energy.

So why don't we have fusion reactors now? Why has it taken so many decades to make progress after the fusion process was mapped out in the 1950s? The problem has been the fiendish difficulty of compressing the hydrogen fuel in a uniform manner. In stars, gravity compresses hydrogen gas into a perfect sphere, so that gas is heated evenly and cleanly.

In NIF's laser fusion, the concentric beams of laser light incinerating the surface of the pellet must be perfectly uniform, and it is exceedingly difficult to achieve this uniformity. In magnetic confinement machines, magnetic fields have both north poles and south poles; as a result, compressing gas evenly into a sphere is extremely difficult. The best we can do is to create a doughnut-shape magnetic field. But compressing the gas is like squeezing a balloon. Every time you squeeze the balloon at one end, air bulges out somewhere else. Squeezing the balloon evenly in all directions simultaneously is a difficult challenge. Hot gas usually leaks out of the magnetic bottle, eventually touching the walls of the reactor and shutting down the fusion process. That is why it has been so hard to squeeze the hydrogen gas for more than about one second.

Unlike the current generation of fission nuclear power plants, a fusion reactor will not create large amounts of nuclear waste. (Each traditional fission plant produces 30 tons of extremely high-level nuclear waste per year. By contrast, the nuclear waste created by a fusion machine would be mainly the radioactive steel left over when the reactor is finally decommissioned.)

Fusion will not completely solve the Earth's energy crisis anytime in the near future; Pierre-Gilles de Gennes, French Nobel laureate in physics, has said, "We say that we will put the sun into a box. The idea is pretty. The problem is, we don't know how to make the box." But if all goes well, researchers are hopeful that within forty years the ITER may pave the way for commercialization of fusion energy, energy that can provide electricity for our homes. One day, fusion reactors may alleviate our energy problem, safely releasing the power of the sun on the Earth.

But even magnetic confinement fusion reactors would not provide enough energy to energize a Death Star weapon. For that we would need an entirely new design.

Nuclear-Fired X-ray Lasers

There is one other possibility for simulating a Death Star laser cannon with today's known technology, and that is with a hydrogen bomb. A battery of X-ray lasers harnessing and focusing the power of nuclear weapons could in theory generate enough energy to operate a device that could incinerate an entire planet.

The nuclear force, pound for pound, releases about 100 million times more energy than a chemical reaction. A piece of enriched uranium no bigger than a baseball is enough to incinerate an entire city in a fiery ball–even though only 1 percent of its mass has been converted to energy. As we discussed, there are a number of ways of injecting energy into a laser beam. By far the most powerful of all is to use the force unleashed by a nuclear bomb.

X-ray lasers have enormous scientific as well as military value. Because of their very short wavelength they can be used to probe atomic distances and decipher the atomic structure of complicated molecules, a feat that is extraordinarily difficult using ordinary methods. A whole new window on chemical reactions opens up when you can "see" the atoms themselves in motion and in their proper arrangement inside a molecule.

Because a hydrogen bomb emits a huge amount of energy in the

X-ray range, X-ray lasers can also be energized by nuclear weapons. The person most closely associated with the X-ray laser is the physicist Edward Teller, father of the hydrogen bomb.

Teller, of course, was the physicist who testified before Congress in the 1950s that Robert Oppenheimer, who had headed the Manhattan Project, could not be trusted to continue work on the hydrogen bomb because of his politics. Teller's testimony led to Oppenheimer's being disgraced and having his security clearance revoked; many prominent physicists never forgave Teller for what he did.

(My own contact with Teller dates from when I was in high school. I conducted a series of experiments on the nature of antimatter and won the grand prize in the San Francisco science fair and a trip to the National Science Fair in Albuquerque, New Mexico. I appeared on local TV with Teller, who was interested in bright young physicists. Eventually I was awarded Teller's Hertz Engineering Scholarship, which paid for my college education at Harvard. I got to know his family fairly well through visits to the Teller household in Berkeley several times a year.)

Basically, Teller's X-ray laser is a small nuclear bomb surrounded by copper rods. The detonation of the nuclear weapon releases a spherical shock wave of intense X-rays. These energetic rays then pass through copper rods, which act as the lasing material, focusing the power of the X-rays into intense beams. These beams of X-rays could then be directed at enemy warheads. Of course, such a device could be used only once, since the nuclear detonation causes the X-ray laser to self-destruct.

The initial test of the nuclear-powered X-ray laser was called the Cabra test, and it took place in 1983 in an underground shaft. A hydrogen bomb was detonated whose flood of incoherent X-rays was then focused into a coherent X-ray laser beam. Initially, the test was deemed a success, and in fact in 1983 it helped to inspire President Ronald Reagan to announce, in a historic speech, his intent to build a "Star Wars" defensive shield. Thus was set in motion a multibillion-dollar effort that continues even to this day to build an array of devices like the nuclear-powered X-ray laser to shoot down enemy ICBMs. (Later investigation showed that the detector used to perform the measurements during the Cabra test was destroyed; hence its readings could not be trusted.)

Can such a controversial device in fact be used today to shoot down ICBM warheads? Perhaps. But an enemy could use a variety of simple, inexpensive methods to nullify such weapons (for example, the enemy could release millions of cheap decoys to fool radar, or spin its warheads to disperse the X-rays, or emit a chemical coating to protect against the X-ray beam). Or an enemy might simply mass-produce warheads to penetrate a Star Wars defensive shield.

So a nuclear-powered X-ray laser today is impractical as a missile defense system. But would it be possible to create a Death Star to use against an approaching asteroid, or to annihilate an entire planet?

The Physics of a Death Star

Can weapons be created that could destroy an entire planet, as in *Star Wars*? In theory, the answer is yes. There are several ways in which they might be created.

First, there is no physical limit to the energy that can be released by a hydrogen bomb. Here's how this works. (The precise outlines of the hydrogen bomb are top secret and classified even today by the U.S. government, but the broad outlines are well known.) A hydrogen bomb is actually built in many stages. By properly stacking these stages in sequence, one could produce a nuclear bomb of almost arbitrary magnitude.

The first stage is a standard fission bomb, using the power of uranium-235 to release a burst of X-rays, as was done in the Hiroshima bomb. In the fraction of a second before the blast from the atomic bomb blows everything apart, the expanding sphere of X-rays outraces the blast (since it travels at the speed of light) and is then refocused onto a container of lithium deuteride, the active substance of a hydrogen bomb. (Precisely how this is done is still classified.) The X-rays striking the lithium deuteride causes it to collapse and heat up to millions of degrees, causing a second explosion, much larger than the first. The burst of X-rays from this hydrogen bomb can then be refocused onto a second piece of lithium deuteride, creating a third explosion. In this way, one could stack lithium deuteride side by side and create a hydrogen bomb of unimaginable magnitude. In fact, the largest hydrogen bomb ever built was a two-stage bomb detonated by the Soviet Union back in 1961, packing the energy of 50 million tons of TNT, although it was theoretically capable of a blast of over 100 million tons of TNT (or about five thousand times the power of the Hiroshima bomb).

To incinerate an entire planet, however, is something of an entirely different magnitude. For this, the Death Star would have to launch thousands of such X-ray lasers into space, and they would then be required to fire all at once. (By comparison, remember that at the height of the cold war the United States and the Soviet Union each accumulated about thirty thousand nuclear bombs.) The collective energy from such an enormous number of X-ray lasers would be enough to incinerate the surface of a planet. So it would certainly be possible for a Galactic Empire hundreds of thousands of years into the future to create such a weapon.

For a very advanced civilization, there is a second option: to create a Death Star using the energy of a gamma ray burster. Such a Death Star would unleash a burst of radiation second only to the big bang. Gamma ray bursters occur naturally in outer space, but it is conceivable that an advanced civilization could harness their vast power. By controlling the spin of a star well before it undergoes a collapse and unleashes a hypernova, one might be able to aim the gamma ray burster at any point in space.

Gamma Ray Bursters

Gamma ray bursters were actually first seen in the 1970s, when the U.S. military launched the *Vela* satellite to detect "nukeflashes" (evidence of an unauthorized detonation of a nuclear bomb). But instead of spotting nukeflashes, the *Vela* satellite detected huge bursts of radiation from space. Initially this discovery set off a panic in the Pentagon: were the Soviets testing a new nuclear weapon in outer space? Later it was determined that these bursts of radiation were coming uniformly from all directions of the sky, meaning that they were actually coming from outside the Milky Way galaxy. But if they were extragalactic, they must be releasing truly astronomical amounts of power, enough to light up the entire visible universe.

When the Soviet Union broke apart in 1990, a huge body of astronomical data was suddenly declassified by the Pentagon, overwhelming astronomers. Suddenly astronomers realized that a new, mysterious phenomenon was staring them in the face, one that would require rewriting the science textbooks.

Since gamma ray bursters last from only a few seconds to a few minutes before they disappear, an elaborate system of sensors is required to spot and analyze them. First, satellites detect the initial burst of radiation and send the exact coordinates of the burster back to Earth. These coordinates are then relayed to optical or radio telescopes, which zero in on the exact location of the gamma ray burster.

Although many details must still be clarified, one theory about the origins of gamma ray bursters is that they are "hypernovae" of enormous strength, which leave massive black holes in their wake. It appears as if gamma ray bursters are monster black holes in formation.

But black holes emit two "jets" of radiation, one from the north pole and one from the south pole, like a spinning top. The radiation seen from a distant gamma ray burster is apparently one of the jets that is aligned toward the Earth. If the jet of a gamma ray burster were aimed at the Earth, and the gamma ray burster were in our galactic neighborhood (a few hundred light-years from Earth), its power would be enough to destroy all life on our planet.

Initially the gamma ray burster's X-ray pulse would create an electromagnetic pulse that would wipe out all electronics equipment on the Earth. Its intense beam of X-rays and gamma rays would be enough to damage the atmosphere of the Earth, destroying our protective ozone layer. The jet of the gamma ray burster would then heat up temperatures on the surface of the Earth, eventually setting off monster firestorms that would engulf the entire planet. The gamma ray burster might not actually explode the entire planet, as in the movie *Star Wars,* but it would certainly destroy all life, leaving a scorched, barren planet.

Conceivably, a civilization hundreds of thousands to a million years more advanced than ours might be able to aim such a black hole in the direction of a target. This could be done by deflecting the path of planets and neutron stars into the dying star at a precise angle just before it collapses. This deflection would be enough to change the spin axis of the star so that it could be aimed in a certain direction. A dying star would make the largest ray gun imaginable.

In summary, the use of powerful lasers to create portable or handheld ray guns and light sabers can be classified as a Class I impossibility–something that is possible in the near future or perhaps within a century. But the extreme challenge of aiming a spinning star before it erupts into a black hole and transforming it into a Death Star would have to be considered a Class II impossibility–something that clearly does not violate the laws of physics (such gamma ray bursters exist) but something that might be possible only thousands to millions of years in the future.

4 : TELEPORTATION

How wonderful that we have met with paradox. Now we have some hope of making progress.

–NIELS BOHR

I canna' change the laws of physics, Captain!

–SCOTTY, CHIEF ENGINEER IN *STAR TREK*

Teleportation, or the ability to transport a person or object instantly from one place to another, is a technology that could change the course of civilization and alter the destiny of nations. It could irrevocably alter the rules of warfare: armies could teleport troops behind enemy lines or simply teleport the enemy's leadership and capture them. Today's transportation system–from cars and ships to airplanes and railroads, and all the many industries that service these systems–would become obsolete; we could simply teleport ourselves to work and our goods to market. Vacations would become effortless, as we teleport ourselves to our destination. Teleportation would change everything.

The earliest mention of teleportation can be found in religious texts such as the Bible, where spirits whisk individuals away. This passage from Acts in the New Testament seems to suggest the teleportation of Philip from Gaza to Azotus: "When they came up out of the water, the Spirit of the Lord suddenly took Philip away, and the eunuch did not see him again, but went on his way rejoicing. Philip, however, appeared at Azotus and traveled about, preaching the gospel in all the towns until he reached Caesarea" (Acts 8:36–40).

Teleportation is also part of every magician's bag of tricks and illusions: pulling rabbits out of a hat, cards out of his or her sleeves, and coins from behind someone's ear. One of the more ambitious magic tricks of recent times featured an elephant disappearing before the eyes of a startled audience. In this demonstration a huge elephant, weighing many tons, was placed inside a cage. Then, with a flick of a magician's wand, the elephant vanished, much to the amazement of the audience. (Of course, the elephant really did not disappear. The trick was performed with mirrors. Long, thin, vertical mirror strips were placed behind each bar of the cage. Like a gate, each of these vertical mirror strips could be made to swivel. At the start of the magic trick, when all these vertical mirror strips were aligned behind the bars, the mirrors could not be seen and the elephant was visible. But when the mirrors were rotated by 45 degrees to face the audience, the elephant disappeared, and the audience was left staring at the reflected image from the side of the cage.)

Teleportation and Science Fiction

The earliest mention of teleportation in science fiction occurred in Edward Page Mitchell's story "The Man Without a Body," published in 1877. In that story a scientist was able to disassemble the atoms of a cat and transmit them over a telegraph wire. Unfortunately the battery died while the scientist was trying to teleport himself. Only his head was successfully teleported.

Sir Arthur Conan Doyle, best known for his Sherlock Holmes novels, was fascinated by the notion of teleportation. After years of writing detective novels and short stories he began to tire of the Sherlock Holmes series and eventually killed off his sleuth, having him plunge to his death with Professor Moriarty over a waterfall. But the public outcry was so great that Doyle was forced to resurrect the detective. Because he couldn't kill off Sherlock Holmes, Doyle instead decided to create an entirely new series, featuring Professor Challenger, who was the counterpart of Sherlock Holmes. Both had a quick wit and a sharp eye for solving mysteries. But while Mr. Holmes used cold, deductive logic to break open complex cases, Professor Challenger explored the dark world of spirituality and paranormal phenomena, including teleportation. In the 1927 novel *The Disintegration Machine,* the professor encountered a gentleman who had invented a machine that could disintegrate a person and then reassemble him somewhere else. But Professor Challenger is horrified when the inventor boasts that his invention could, in the wrong hands, disintegrate entire cities with millions of people with a push of a button. Professor Challenger then uses the machine to disintegrate the inventor, and leaves the laboratory, without reassembling him.

More recently Hollywood has discovered teleportation. The 1958 film *The Fly* graphically examined what could happen when teleportation goes horribly awry. When a scientist successfully teleports himself across a room, his atoms mix with those of a fly that accidentally entered the teleportation chamber, so the scientist turns into a grotesquely mutated monster, half human and half fly. (A remake featuring Jeff Goldblum was released in 1986.)

Teleportation first became prominent in popular culture with the *Star Trek* series. Gene Roddenberry, *Star Trek*'s creator, introduced teleportation into the series because the Paramount Studio budget did not allow for the costly special effects needed to simulate rocket ships taking off and landing on distant planets. It was cheaper simply to beam the crew of the *Enterprise* to their destination.

Over the years any number of objections have been raised by scientists about the possibility of teleportation. To teleport someone, you would have to know the precise location of every atom in a living body, which would probably violate the Heisenberg uncertainty principle (which states that you cannot know both the precise location and the velocity of an electron). The producers of the *Star Trek* series, bowing to the critics, introduced "Heisenberg Compensators" in the transporter room, as if one could compensate for the laws of quantum physics by adding a gadget to the transporter. But as it turns out, the need to create these Heisenberg Compensators might have been premature. Early critics and scientists may have been wrong.

Teleportation and the Quantum Theory

According to Newtonian theory, teleportation is clearly impossible. Newton's laws are based on the idea that matter is made of tiny, hard billiard balls. Objects do not move until they are pushed; objects do not suddenly disappear and reappear somewhere else.

But in the quantum theory, that's precisely what particles can do. Newton's laws, which held sway for 250 years, were overthrown in 1925 when Werner Heisenberg, Erwin Schrödinger, and their colleagues developed the quantum theory. When analyzing the bizarre properties of atoms, physicists discovered that electrons acted like waves and could make quantum leaps in their seemingly chaotic motions within the atom.

The man most closely associated with these quantum waves is the Viennese physicist Erwin Schrödinger, who wrote down the celebrated wave equation that bears his name, one of the most important in all of physics and chemistry. Entire courses in graduate school are devoted to solving his famous equation, and entire walls of physics libraries are full of books that examine its profound consequences. In principle, the sum total of all of chemistry can be reduced to solutions to this equation.

In 1905 Einstein had shown that waves of light can have particle-like properties; that is, they can be described as packets of energy called photons. But by the 1920s it was becoming apparent to Schrödinger that the opposite was also true: that particles like electrons could exhibit wavelike behavior. This idea was first pointed out by French physicist Louis de Broglie, who won the Nobel Prize for this conjecture. (We demonstrate this to our undergraduate students at our university. We fire electrons inside a cathode ray tube, like those commonly found in TVs. The electrons pass through a tiny hole, so normally you would expect to see a tiny dot where the electrons hit the TV screen. Instead you find concentric, wavelike rings, which you would expect if a wave had passed through the hole, not a point particle.)

One day Schrödinger gave a lecture on this curious phenomenon. He was challenged by a fellow physicist, Peter Debye, who asked him: If electrons are described by waves, then what is their wave equation?

Ever since Newton created the calculus, physicists had been able to describe waves in terms of differential equations, so Schrödinger took Debye's question as a challenge to write down the differential equation for electron waves. That month Schrödinger went on vacation, and when he came back he had that equation. So in the same way that Maxwell before him had taken the force fields of Faraday and extracted Maxwell's equations for light, Schrödinger took the matter-waves of de Broglie and extracted Schrodinger's equations for electrons.

(Historians of science have spent some effort trying to track down precisely what Schrödinger was doing when he discovered his celebrated equation that forever changed the landscape of modern physics and chemistry. Apparently, Schrödinger was a believer in free love and would often be accompanied on vacation by his mistresses and his wife. He even kept a detailed diary account of all his numerous lovers, with elaborate codes concerning each encounter. Historians now believe that he was in the Villa Herwig in the Alps with one of his girlfriends the weekend that he discovered his equation.)

When Schrödinger began to solve his equation for the hydrogen atom, he found, much to his surprise, the precise energy levels of hydrogen that had been carefully catalogued by previous physicists. He then realized that the old picture of the atom by Niels Bohr showing electrons whizzing around the nucleus (which is used even today in books and advertisements when trying to symbolize modern science) was actually wrong. These orbits would have to be replaced by waves surrounding the nucleus.

Schrödinger's work sent shock waves, as well, through the physics community. Suddenly physicists were able to peer inside the atom itself, to examine in detail the waves that made up its electron shells, and to extract precise predictions for these energy levels that fit the data perfectly.

But there was still a nagging question that haunts physics even today. If the electron is described by a wave, then what is waving? This has been answered by physicist Max Born, who said that these waves are actually waves of probability. These waves tell you only the chance of finding a particular electron at any place and any time. In other words, *the electron is a particle, but the probability of finding that particle is given by Schrödinger's wave.* The larger the wave, the greater the chance of finding the particle at that point.

With these developments, suddenly chance and probability were being introduced right into the heart of physics, which previously had given us precise predictions and detailed trajectories of particles, from planets to comets to cannon balls.

This uncertainty was finally codified by Heisenberg when he proposed the uncertainty principle, that is, the concept that you cannot know both the exact velocity and the position of an electron at the same time. Nor can you know its exact energy, measured over a given amount of time. At the quantum level all the basic laws of common sense are violated: electrons can disappear and reappear elsewhere, and electrons can be many places at the same time.

(Ironically, Einstein, the godfather of the quantum theory who helped to start the revolution in 1905, and Schrödinger, who gave us the wave equation, were horrified by the introduction of chance into fundamental physics. Einstein wrote, "Quantum mechanics calls for a great deal of respect. But some inner voice tells me that this is not the true Jacob. The theory offers a lot, but it hardly brings us any closer to the Old Man's secret. For my part, at least, I am convinced that He doesn't throw dice.")

Heisenberg's theory was revolutionary and controversial–but it worked. In one sweep, physicists could explain a vast number of puzzling phenomena, including the laws of chemistry. To impress my Ph.D. students with just how bizarre the quantum theory is, I sometimes ask them to calculate the probability that their atoms will suddenly dissolve and reappear on the other side of a brick wall. Such a teleportation event is impossible under Newtonian physics but is actually allowed under quantum mechanics. The answer, however, is that one would have to wait longer than the lifetime of the universe for this to occur. (If you used a computer to graph the Schrödinger wave of your own body, you would find that it very much resembles all the features of your body, except that the graph would be a bit fuzzy, with some of your waves oozing out in all directions. Some of your waves would extend even as far as the distant stars. So there is a very tiny probability that one day you might wake up on a distant planet.)

The fact that electrons can seemingly be many places at the same time forms the very basis of chemistry. We know that electrons circle around the nucleus of an atom, like a miniature solar system. But atoms and solar systems are quite different; if two solar systems collide in outer space, the solar systems break apart and planets are flung into deep space. Yet when atoms collide they often form molecules that are perfectly stable, sharing electrons between them. In high school chemistry class the teacher often represents this with a "smeared electron," which resembles a football, connecting the two atoms together.

But what chemistry teachers rarely tell their students is that the electron is not "smeared" between two atoms at all. This "football" actually represents the probability that the electron is in many places at the same time within the football. In other words, all of chemistry, which explains the molecules inside our bodies, is based on the idea that electrons can be many places at the same time, and it is this sharing of electrons between two atoms that holds the molecules of our body together. *Without the quantum theory, our molecules and atoms would dissolve instantly.*

This peculiar but profound property of the quantum theory (that there is a finite probability that even the most bizarre events may happen) was exploited by Douglas Adams in his hilarious novel *The Hitchhiker's Guide to the Galaxy.* He needed a convenient way to whiz through the galaxy, so he invented the Infinite Improbability Drive, "a wonderful new method of crossing vast interstellar distances in a mere nothingth of a second, without all that tedious mucking around in hyperspace." His machine enables you to change the odds of any quantum event at will, so that even highly improbable events become commonplace. So if you want to jet off to the nearest star system, you would simply change the probability that you will rematerialize on that star, and voilà! You would be instantly teleported there.

In reality the quantum "jumps" so common inside the atom cannot be easily generalized to large objects such as people, which contain trillions upon trillions of atoms. Even if the electrons in our body are dancing and jumping in their fantastic journey around the nucleus, there are so many of them that their motions average out. That is, roughly speaking, why at our level substances seem solid and permanent.

So while teleportation is allowed at the atomic level, one would have to wait longer than the lifetime of the universe to actually witness these bizarre effects on a macroscopic scale. But can one use the laws of the quantum theory to create a machine to teleport something on demand, as in science fiction stories? Surprisingly, the answer is a qualified yes.

The EPR Experiment

The key to quantum teleportation lies in a celebrated 1935 paper by Albert Einstein and his colleagues Boris Podolsky and Nathan Rosen, who, ironically, proposed the EPR experiment (named for the three authors) to kill off, once and for all, the introduction of probability into physics. (Bemoaning the undeniable experimental successes of the quantum theory, Einstein wrote, "the more success the quantum theory has, the sillier it looks.")

If two electrons are initially vibrating in unison (a state called coherence) they can remain in wavelike synchronization even if they are separated by a large distance. Although the two electrons may be separated by light-years, there is still an invisible Schrödinger wave connecting both of them, like an umbilical cord. If something happens to one electron, then some of that information is immediately transmitted to the other. This is called "quantum entanglement," the concept that particles vibrating in coherence have some kind of deep connection linking them together.

Let's start with two coherent electrons oscillating in unison. Next, let them go flying out in opposite directions. Each electron is like a spinning top. The spins of each electron can be pointed up or down. Let's say that the total spin of the system is zero, so that if the spin of one electron is up, then you know automatically that the spin of the other electron is down. According to the quantum theory, before you make a measurement, the electron is spinning neither up nor down but exists in a nether state where it is spinning both up and down simultaneously. (Once you make an observation, the wave function "collapses," leaving a particle in a definite state.)

Next, measure the spin of one electron. It is, say, spinning up. Then you know instantly that the spin of the other electron is down. Even if the electrons are separated by many light-years, you instantly know the spin of the second electron as soon as you measure the spin of the first electron. In fact, *you know this faster than the speed of light*! Because these two electrons are "entangled," that is, their wave functions beat in unison, their wave functions are connected by an invisible "thread" or umbilical cord. Whatever happens to one automatically has an effect on the other. (This means, in some sense, that what happens to us automatically affects things instantaneously in distant corners of the universe, since our wave functions were probably entangled at the beginning of time. In some sense there is a web of entanglement that connects distant corners of the universe, including us.) Einstein derisively called this "spooky-action-at-distance," and this phenomenon enabled him to "prove" that the quantum theory was wrong, in his mind, since nothing can travel faster than the speed of light.

Originally, Einstein designed the EPR experiment to serve as the death knell of the quantum theory. But in the 1980s Alan Aspect and his colleagues in France performed this experiment with two detectors separated by 13 meters, measuring the spins of photons emitted from calcium atoms, and the results agreed precisely with the quantum theory. Apparently God does play dice with the universe.

Did information really travel faster than light? Was Einstein wrong about the speed of light being the speed limit of the universe? Not really. Information did travel faster than the speed of light, but the information was random, and hence useless. You cannot send a real message, or Morse code, via the EPR experiment even if information is traveling faster than light.

Knowing that an electron on the other side of the universe is spinning down is useless information. You cannot send today's stock quotations via this method. For example, let's say that a friend always wears one red and one green sock, in random order. Let's say you examine one leg, and the leg has a red sock on it. Then you know, faster than the speed of light, that the other sock is green. Information actually traveled faster than light, but this information is useless. No signal containing nonrandom information can be sent via this method.

For years the EPR experiment was used as an example of the resounding victory of the quantum theory over its critics, but it was a hollow victory with no practical consequences. Until now.

Quantum Teleportation

Everything changed in 1993, when scientists at IBM, led by Charles Bennett, showed that it was physically possible to teleport objects, at least at the atomic level, using the EPR experiment. (More precisely, they showed that you could teleport all the information contained within a particle.) Since then physicists have been able to teleport photons and even entire cesium atoms. Within a few decades scientists may be able to teleport the first DNA molecule and virus.

Quantum teleportation exploits some of the more bizarre properties of the EPR experiment. In these teleportation experiments physicists start with two atoms, A and C. Let's say we wish to teleport information from atom A to atom C. We begin by introducing a third atom, B, which starts out being entangled with C, so B and C are coherent. Now atom A comes in contact with atom B. A scans B, so that the information content of atom A is transferred to atom B. A and B become entangled in the process. But since B and C were originally entangled, the information within A has now been transferred to atom C. In conclusion, atom A has now been teleported into atom C, that is, the information content of A is now identical to that of C.

Notice that the information within atom A has been destroyed (so we don't have two copies after the teleportation). This means that anyone being hypothetically teleported would die in the process. But the information content of his body would appear elsewhere. Notice also that atom A did not move to the position of atom C. On the contrary, it is the information within A (e.g., its spin and polarization) that has been transferred to C. (This does not mean that atom A was dissolved and then zapped to another location. It means that the information content of atom A has been transferred to another atom, C.)

Since the original announcement of this breakthrough, progress has been fiercely competitive as different groups have attempted to outrace each other. The first historic demonstration of quantum teleportation in which photons of ultraviolet light were teleported occurred in 1997 at the University of Innsbruck. This was followed the next year by experimenters at Cal Tech who did an even more precise experiment involving teleporting photons.

In 2004 physicists at the University of Vienna were able to teleport particles of light over a distance of 600 meters beneath the River Danube, using a fiber-optic cable, setting a new record. (The cable itself was 800 meters long and was strung underneath the public sewer system beneath the River Danube. The sender stood on one side of the river, and the receiver was on the other.)

One criticism of these experiments is that they were conducted with photons of light. This is hardly the stuff of science fiction. It was significant, therefore, in 2004, when quantum teleportation was demonstrated not with photons of light, but with actual atoms, bringing us a step closer to a more realistic teleportation device. The physicists at the National Institute of Standards and Technology in Washington, D.C., successfully entangled three beryllium atoms and transferred the properties of one atom into another. This achievement was so significant that it made the cover of *Nature* magazine. Another group was able to teleport calcium atoms as well.

In 2006 yet another spectacular advance was made, for the first time involving a macroscopic object. Physicists at the Niels Bohr Institute in Copenhagen and the Max Planck Institute in Germany were able to entangle a light beam with a gas of cesium atoms, a feat involving trillions upon trillions of atoms. Then they encoded information contained inside laser pulses and were able to teleport this information to the cesium atoms over a distance of about half a yard. "For the first time," said Eugene Polzik, one of the researchers, quantum teleportation "has been achieved between light–the carrier of information–and atoms."

Teleportation Without Entanglement

Progress in teleportation is rapidly accelerating. In 2007 yet another breakthrough was made. Physicists proposed a teleportation method that does not require entanglement. We recall that entanglement is the single most difficult feature of quantum teleportation. Solving this problem could open up new vistas in teleportation.

"We're talking about a beam of about 5,000 particles disappearing from one place and appearing somewhere else," says physicist Aston Bradley of the Australian Research Council Centre of Excellence for Quantum Atom Optics in Brisbane, Australia, who helped pioneer a new method of teleportation.

"We feel that our scheme is closer in spirit to the original fictional concept," he claims. In their approach, he and his colleagues take a beam of rubidium atoms, convert all its information into a beam of light, send this beam of light across a fiber-optic cable, and then reconstruct the original beam of atoms in a distant location. If his claim holds up, this method would eliminate the number one stumbling block to teleportation and open up entirely new ways to teleport increasingly large objects.

In order to distinguish this new method from quantum teleportation, Dr. Bradley has called his method "classical teleportation." (This is a bit misleading, since his method also depends heavily on the quantum theory, but not on entanglement.)

The key to this novel type of teleportation is a new state of matter called a "Bose Einstein condensate," or BEC, which is one of the coldest substances in the entire universe. In nature the coldest temperature is found in outer space; it is 3 K above absolute zero. (This is due to residual heat left over from the big bang, which still fills up the universe.) But a BEC is a *millionth to a billionth* of a degree above absolute zero, a temperature that can be found only in the laboratory.

When certain forms of matter are cooled down to near absolute zero, their atoms all tumble down to the lowest energy state, so that all their atoms vibrate in unison, becoming coherent. The wave functions of all the atoms overlap, so that, in some sense, a BEC is like a gigantic "super atom," with all the individual atoms vibrating in unison. This bizarre state of matter was predicted by Einstein and Satyendranath Bose in 1925, but it would be another seventy years, not until 1995, before a BEC was finally created in the lab at MIT and the University of Colorado.

Here's how Bradley and company's teleportation device works. First they start with a collection of supercold rubidium atoms in a BEC state. They then apply a beam of matter to the BEC (also made of rubidium atoms). These atoms in the beam also want to tumble down to the lowest energy state, so they shed their excess energy in the form of a pulse of light. This light beam is then sent down a fiber-optic cable. Remarkably the light beam contains all the quantum information necessary to describe the original matter beam (e.g., the location and velocity of all its atoms). Then the light beam hits another BEC, which then converts the light beam into the original matter beam.

This new teleportation method has tremendous promise, since it doesn't involve the entanglement of atoms. But this method also has its problems. It depends crucially on the properties of BECs, which are difficult to create in the laboratory. Furthermore, the properties of BECs are quite peculiar, because they behave as if they were one gigantic atom. In principle, bizarre quantum effects that we see only at the atomic level can be seen with the naked eye with a BEC. This was once thought to be impossible.

The immediate practical application of BECs is to create "atomic lasers." Lasers, of course, are based on coherent beams of photons vibrating in unison. But a BEC is a collection of atoms vibrating in unison, so it's possible to create beams of BEC atoms that are all coherent. In other words, a BEC can create the counterpart of the laser, the atomic laser or matter laser, which is made of BEC atoms. The commercial applications of lasers are enormous, and the commercial applications of atomic lasers could also be just as profound. But because BECs exist only at temperatures hovering just above absolute zero, progress in this field will be slow, albeit steady.

Given the progress we have made, when might we be able to teleport ourselves? Physicists hope to teleport complex molecules in the coming years. After that perhaps a DNA molecule or even a virus may be teleported within decades. There is nothing in principle to prevent teleporting an actual person, just as in the science fiction movies, but the technical problems facing such a feat are truly staggering. It takes some of the finest physics laboratories in the world just to create coherence between tiny photons of light and individual atoms. Creating quantum coherence involving truly macroscopic objects, such as a person, is out of the question for a long time to come. In fact, it will likely take many centuries, or longer, before everyday objects could be teleported–if it's possible at all.

Quantum Computers

Ultimately, the fate of quantum teleportation is intimately linked to the fate of the development of quantum computers. Both use the same quantum physics and the same technology, so there is intense cross-fertilization between these two fields. Quantum computers may one day replace the familiar digital computer sitting on our desks. In fact, the future of the world's economy may one day depend on such computers, so there is enormous commercial interest in these technologies. One day Silicon Valley could become a Rust Belt, replaced by new technologies emerging from quantum computing.

Ordinary computers compute on a binary system of 0s and 1s, called bits. But quantum computers are far more powerful. They can compute on qubits, which can take values between 0 and 1. Think of an atom placed in a magnetic field. It is spinning like a top, so its spin axis can point either up or down. Common sense tells us that the spin of the atom can be either up or down but not both at the same time. But in the strange world of the quantum, the atom is described as the sum of two states, the sum of an atom spinning up and an atom spinning down. In the netherworld of the quantum, every object is described by the sum of all possible states. (If large objects, like cats, are described in this quantum fashion, it means that you have to add the wave function of a live cat to that of a dead cat, so the cat is neither dead nor alive, as I will discuss in greater detail in Chapter 13.)

Now imagine a string of atoms aligned in a magnetic field, with the spin aligned in one fashion. If a laser beam is shone on this string of atoms the laser beam will bounce off this collection of atoms, flipping the spin axis of some of the atoms. By measuring the difference between the incoming and outgoing laser beam, we have accomplished a complicated quantum "calculation," involving the flipping of many spins.

Quantum computers are still in their infancy. The world's record for a quantum computation is $3 \times 5 = 15$, hardly a calculation that will supplant today's supercomputers. Quantum teleportation and quantum computers both share the same fatal weakness: maintaining coherence for large collections of atoms. If this problem can be solved, it would be an enormous breakthrough in both fields.

The CIA and other secret organizations are intensely interested in quantum computers. Many of the world's secret codes depend on a "key," which is a very large integer, and one's ability to factor it into prime numbers. If the key is the product of two numbers, each with one hundred digits, then it might take a digital computer more than a hundred years to find these two factors from scratch. Such a code is essentially unbreakable today.

But in 1994 Peter Shor of Bell Labs showed that factoring large numbers could be child's play for a quantum computer. This discovery immediately piqued the interest of the intelligence community. In principle a quantum computer could break all the world's codes, throwing the security of today's computer systems into total disorder. The first country that is able to build such a system would be able to unlock the deepest secrets of other nations and organizations.

Some scientists have speculated that in the future the world's economy might depend on quantum computers. Silicon-based digital computers are expected to reach their physical limits in terms of increased computer power sometime after 2020. A new, more powerful family of computers might be necessary if technology is going to continue to advance. Others are exploring the possibility of reproducing the power of the human brain via quantum computers.

The stakes, therefore, are very high. If we can solve the problem of coherence, not only might we be able to solve the challenge of teleportation; we might also have the ability to advance technology of all kinds in untold ways via quantum computers. This breakthrough is so important that I will return to this discussion in later chapters.

As I pointed out earlier, coherence is extraordinarily difficult to maintain in the lab. The tiniest vibration could upset the coherence of two atoms and destroy the computation. Today it is very difficult to maintain coherence in more than just a handful of atoms. Atoms that are originally in phase begin to decohere within a matter of nanoseconds to, at best, a second. Teleportation must be done very rapidly, before the atoms begin to decohere, thus placing another restriction on quantum computation and teleportation.

In spite of these challenges, David Deutsch of Oxford University believes that these problems can be overcome: "With luck, and with the help of recent theoretical advances, [a quantum computer] may take a lot less than 50 years. . . . It would be an entirely new way of harnessing nature."

To build a useful quantum computer we would need to have hundreds to millions of atoms vibrating in unison, an achievement far beyond our capabilities today. Teleporting Captain Kirk would be astronomically difficult. We would have to create a quantum entanglement with a twin of Captain Kirk. Even with nanotechnology and advanced computers, it is difficult to see how this could be accomplished.

So teleportation exists at the atomic level, and we may eventually teleport complex and even organic molecules within a few decades. But the teleportation of a macroscopic object will have to wait for several decades to centuries beyond that, or longer, if indeed it is even possible. Therefore teleporting complex molecules, perhaps even a virus or a living cell, qualifies as a Class I impossibility, one that should be possible within this century. But teleporting a human being, although it is allowed by the laws of physics, may take many centuries beyond that, assuming it is possible at all. Hence I would qualify that kind of teleportation as a Class II impossibility.

5 : TELEPATHY

If you haven't found something strange during the day,
it hasn't been much of a day.

—JOHN WHEELER

Only those who attempt the absurd will achieve the impossible.

—M. C. ESCHER

A. E. van Vogt's novel *Slan* captures the vast potential and our darkest fears associated with the power of telepathy.

Jommy Cross, the protagonist in the novel, is a "slan," a dying race of superintelligent telepaths.

His parents were brutally murdered by enraged mobs of humans, who fear and despise all telepaths, because of the enormous power wielded by those who can intrude on their private, most intimate thoughts. Humans mercilessly hunt down the slans like animals. With their characteristic tendrils growing out of their heads, slans are easy to spot. In the course of the book, Jommy tries to make contact with other slans who might have fled into outer space to escape the witch hunts of humans determined to exterminate them.

Historically, mind reading has been seen as so important that it has often been associated with the gods. One of the most fundamental powers of any god is the ability to read our minds and hence answer our deepest prayers. A true telepath who could read minds at will could easily become the wealthiest, most powerful person on Earth, able to enter into the minds of Wall Street bankers or to blackmail and coerce his rivals. He would pose a threat to the security of governments. He could effortlessly steal a nation's most sensitive secrets. Like the slans, he would be feared and perhaps hunted down.

The enormous power of a true telepath was highlighted in the landmark Foundation series by Isaac Asimov, often touted as one of the greatest science fiction epics of all time. A Galactic Empire that has ruled for thousands of years is on the verge of collapse and ruin. A secret society of scientists, called the Second Foundation, uses complex equations to predict that the Empire will eventually fall and plunge civilization into thirty thousand years of darkness. The scientists draft an elaborate plan based on their equations in an effort to reduce this collapse of civilization down to just a few thousand years. But then disaster strikes. Their elaborate equations fail to predict a single event, the birth of a mutant called the Mule, who is capable of controlling minds over great distances and hence able to seize control of the Galactic Empire. The galaxy is doomed to thirty thousand years of chaos and anarchy unless this telepath can be stopped.

Although science fiction is full of fantastic tales concerning telepaths, the reality is much more mundane. Because thoughts are private and invisible, for centuries charlatans and swindlers have taken advantage of the naïve and gullible among us. One simple parlor trick used by magicians and mentalists is to use a shill–an accomplice planted in the audience whose mind is then "read" by the mentalist.

The careers of several magicians and mentalists, in fact, have been based on the famous "hat trick," in which people write private messages on strips of paper, which are then placed in a hat. The magician then proceeds to tell the audience what is written on each strip of paper, amazing everyone. There is a deceptively simple explanation for this ingenious trick (see the notes).

One of the most famous cases of telepathy did not involve a shill but an animal, Clever Hans, a wonder horse that astonished European audiences in the 1890s. Clever Hans, to the amazement of audiences, could perform complex mathematical feats of calculation. If, for example, you asked Clever Hans to divide 48 by 6, the horse would beat its hoof 8 times. Clever Hans, in fact, could divide, multiply, add fractions, spell, and even identify musical tones. Clever Hans's fans declared that he was either more intelligent than many humans, or he could telepathically pick people's brains.

But Clever Hans was not the product of some clever trickery. The marvelous ability of Clever Hans to perform arithmetic even fooled his trainer. In 1904 prominent psychologist Professor C. Strumpf was brought in to analyze the horse and could find no obvious evidence of trickery or covert signaling to the horse, only adding to the public's fascination with Clever Hans. Three years later, however, Strumpf's student, psychologist Oskar Pfungst, did much more rigorous testing and finally discovered Clever Hans's secret. All he really did was observe the subtle facial expressions of his trainer. The horse would continue to beat his hoofs until his trainer's facial expression changed slightly, at which point he would stop beating. Clever Hans could not read people's minds or perform arithmetic; he was simply a shrewd observer of people's faces.

There have been other "telepathic" animals in recorded history. As early as 1591 a horse named Morocco became famous in England and made a fortune for his owner by picking out people in the audience, pointing out letters of the alphabet, and adding the total of a pair of dice. He caused such a sensation in England that Shakespeare immortalized him in his play *Love's Labour's Lost* as "the dancing horse."

Gamblers also are able to read people's minds in a limited sense. When a person sees something pleasurable, the pupils of his eyes usually dilate. When he sees something undesirable (or performs a mathematical calculation), his pupils contract. Gamblers can read the emotions of their poker-faced opponents by looking for their eyes to dilate or contract. This is one reason that gamblers often wear colored visors over their eyes, to shield their pupils. One can also bounce a laser beam off a person's pupil and analyze where it is reflected, and thereby determine precisely where a person is looking. By analyzing the motion of the reflected dot of laser light, one can determine how a person scans a picture. By combining these two technologies, one can then determine a person's emotional reaction as he scans a picture, all without his permission.

PSYCHICAL RESEARCH

The first scientific studies of telepathy and other paranormal phenomenon were conducted by the Society for Psychical Research, founded in London in 1882. (The term "mental telepathy" was coined that year by F. W. Myers, an associate of the society.) Past presidents of this society included some of the most notable figures of the nineteenth century. The society, which still exists today, was able to debunk the claims of many frauds, but was often split between the spiritualists, who firmly believed in the paranormal, and the scientists, who wanted more serious scientific study.

One researcher connected with the society, Dr. Joseph Banks Rhine, began the first systematic and rigorous study of psychic phenomena in the United States in 1927, founding the Rhine Institute (now called the Rhine Research Center) at Duke University, North Carolina. For decades he and his wife, Louisa, conducted some of the first scientifically controlled experiments in the United States on a wide variety of parapsychological phenomena and published them in peer-reviewed publications. It was Rhine who coined the term "extrasensory perception" (ESP) in one of his first books.

Rhine's laboratory, in fact, set the standard for psychic research. One of his associates, Dr. Karl Zener, developed the five-symbol card system, now known as Zener cards, for analyzing telepathic powers. The vast majority of results showed absolutely no evidence of telepathy. But a small minority of experiments seemed to show small but remarkable correlations in the data that could not be explained by pure chance. The problem was that these experiments often could not be duplicated by other researchers.

Although Rhine tried to establish a reputation for rigor, his reputation was somewhat tarnished by an encounter with a horse called

Lady Wonder. This horse could perform dazzling feats of telepathy, such as knocking over toy alphabet blocks and thereby spelling out words that members of an audience were thinking. Rhine apparently did not know about the Clever Hans effect. In 1927 Rhine analyzed Lady Wonder in some detail and concluded, "There is left then, only the telepathic explanation, the transference of mental influence by an unknown process. Nothing was discovered that failed to accord with it, and no other hypothesis proposed seems tenable in view of the results." Later, Milbourne Christopher revealed the true source of Lady Wonder's telepathic power: subtle motions of the whip carried by the horse's owner. The subtle movements of the whip were cues for Lady Wonder to stop beating her hoof. (But even after the true source of Lady Wonder's power was exposed, Rhine continued to believe that the horse was truly telepathic, but somehow had lost its telepathic power, forcing the owner to resort to trickery.)

Rhine's reputation suffered a final crushing blow, however, when he was on the verge of retiring. He was seeking a successor with an untarnished reputation to carry on the work of his institute. One promising candidate was Dr. Walter Levy, whom he hired in 1973. Dr. Levy was a rising star in the field, reporting sensational study results seeming to demonstrate that mice could telepathically alter a computer's random number generator. However, suspicious lab workers discovered that Dr. Levy was surreptitiously sneaking into the lab at night to alter the results of the tests. He was caught red-handed doctoring the data. Further tests showed that the mice possessed no telepathic power whatsoever, and Dr. Levy was forced to resign from the institute in disgrace.

Telepathy and Star Gate

Interest in the paranormal took a deadly turn at the height of the cold war, during which a number of clandestine experiments on telepathy, mind control, and remote viewing were spawned. (Remote viewing is "seeing" a distant location by the mind alone, by reading the minds of others.) Star Gate was the code name for a number of secret CIA-sponsored studies (such as Sun Streak, Grill Flame, and Center Lane). The efforts got their start around 1970 when the CIA concluded that the Soviet Union was spending up to 60 million rubles a year on "psychotronic" research. There was concern that the Soviets might be using ESP to locate U.S. submarines and military installations, to identify spies, and to read secret papers.

Funding for the CIA studies began in 1972, and Russell Targ and Harold Puthoff of the Stanford Research Institute (SRI) in Menlo Park were in charge. Initially, they sought to train a cadre of psychics who could engage in "psychic warfare." Over more than two decades, the United States spent $20 million on Star Gate, with over forty personnel, twenty-three remote viewers, and three psychics on the payroll.

By 1995, with a budget of $500,000 per year, the CIA had conducted hundreds of intelligence-gathering projects involving thousands of remote viewing sessions. Specifically, the remote viewers were asked to

- locate Colonel Gadhafi before the 1986 bombing of Libya
- find plutonium stockpiles in North Korea in 1994
- locate a hostage kidnapped by the Red Brigades in Italy in 1981
- locate a Soviet Tu-95 bomber that had crashed in Africa

In 1995, the CIA asked the American Institute for Research (AIR) to evaluate these programs. The AIR recommended that the programs be shut down. "There's no documented evidence it had any value to the intelligence community," wrote David Goslin of the AIR.

Proponents of Star Gate boasted that over the years they had scored "eight-martini" results (conclusions that were so spectacular that you had to go out and drink eight martinis to recover). Critics, however, maintained that a huge majority of the remote viewing produced worthless, irrelevant information, wasting taxpayer dollars, and that the few "hits" they scored were vague and so general that they could be applied to any number of situations. The AIR report stated that the most impressive "successes" of Star Gate involved remote viewers who already had some knowledge of the operation they were studying, and hence might have made educated guesses that sounded reasonable.

In the end the CIA concluded that Star Gate had yielded not a single instance of information that helped the agency guide intelligence operations, so it canceled the project. (Rumors persisted that the CIA used remote viewers to locate Saddam Hussein during the Gulf War, although all efforts were unsuccessful.)

Brain Scans

At the same time, scientists were beginning to understand some of the physics behind the workings of the brain. In the nineteenth century scientists suspected that electrical signals were being transmitted inside the brain. In 1875 Richard Caton discovered that by placing electrodes on the surface of the head it was possible to detect the tiny electrical signals emitted by the brain. This eventually led to the invention of the electroencephalograph (EEG).

In principle the brain is a transmitter over which our thoughts are broadcast in the form of tiny electrical signals and electromagnetic waves. But there are problems with using these signals to read someone's thoughts. First, the signals are extremely weak, in the milliwatt range. Second, the signals are gibberish, largely indistinguishable from random noise. Only crude information about our thoughts can be gleaned from this garble. Third, our brain is not capable of receiving similar messages from other brains via these signals; that is, we lack an antenna. And, finally, even if we could receive these faint signals, we could not unscramble them. Using ordinary Newtonian and Maxwellian physics, telepathy via radio does not seem to be possible.

Some believe that perhaps telepathy is mediated by a fifth force, called the "psi" force. But even advocates of parapsychology admit that they have no concrete, reproducible evidence of this psi force.

But this leaves open the question: What about telepathy using the quantum theory?

In the last decade, new quantum instruments have been introduced that for the first time in history enable us to look into the thinking brain.

Leading this quantum revolution are the PET (positron-emission tomography) and MRI (magnetic resonance imaging) brain scans. A PET scan is created by injecting radioactive sugar into the blood. This sugar concentrates in parts of the brain that are activated by the thinking process, which requires energy. The radioactive sugar emits positrons (antielectrons) that are easily detected by instruments. Thus, by tracing out the pattern created by antimatter in the living brain, one can also trace out the patterns of thought, isolating precisely which parts of the brain are engaged in which activity.

The MRI machine operates in the same way, except it is more precise. A patient's head is placed inside a huge doughnut-shaped magnetic field. The magnetic field makes the nuclei of the atoms in the brain align parallel to the field lines. A radio pulse is sent into the patient, making these nuclei wobble. When the nuclei flip orientation, they emit a tiny radio "echo" that can be detected, thereby signaling the presence of a particular substance. For example, brain activity is related to oxygen consumption, so the MRI machine can isolate the process of thinking by zeroing in on the presence of oxygenated blood. The higher the concentration of oxygenated blood, the greater the mental activity in that part of the brain. (Today "functional MRI machines" [fMRI] can zero in on tiny areas of the brain only a millimeter across in fractions of a second, making these machines ideal for tracing out the pattern of thoughts of the living brain.)

MRI Lie Detectors

With MRI machines, there is a possibility that one day scientists may be able to decipher the broad outlines of thoughts in the living brain. The simplest test of "mind reading" would be to determine whether or not someone is lying.

According to legend, the world's first lie detector was created by an Indian priest centuries ago. He would put the suspect and a "magic donkey" into a sealed room, with the instruction that the suspect should pull on the magic donkey's tail. If the donkey began to talk, it meant the suspect was a liar. If the donkey remained silent, then the suspect was telling the truth. (But secretly, the elder would put soot on the donkey's tail.)

After the suspect was taken out of the room, the suspect would usually proclaim his innocence because the donkey did not speak when he pulled its tail. But the priest would then examine the suspect's hands. If the hands were clean, it meant he was lying. (Sometimes the threat of using a lie detector is more effective than the lie detector itself.)

The first "magic donkey" in modern times was created in 1913, when psychologist William Marston wrote about analyzing a person's blood pressure, which would be elevated when telling a lie. (This observation about blood pressure actually goes back to ancient times, when a suspect would be questioned while an investigator held on to his hands.) The idea soon caught on, and soon even the Department of Defense was setting up its own Polygraph Institute.

But over the years it has become clear that lie detectors can be fooled by sociopaths who show no remorse for their actions. The most famous case was that of the CIA double agent Aldrich Ames, who pocketed huge sums of money from the former Soviet Union by sending scores of U.S. agents to their death and divulging secrets of the U.S. nuclear navy. For decades Ames sailed through a battery of the CIA's lie detector tests. So, too, did serial killer Gary Ridgway, known as the notorious Green River Killer; he killed as many as fifty women.

In 2003 the U.S. National Academy of Sciences issued a scathing report on the reliability of lie detectors, listing all the ways in which lie detectors could be fooled and innocent people branded as liars.

But if lie detectors measure only anxiety levels, what about measuring the brain itself? The idea of looking into brain activity to ferret out lies dates back twenty years, to the work of Peter Rosenfeld of Northwestern University, who observed that EEG scans of people in the process of lying showed a different pattern in the P300 waves than when those people were telling the truth. (P300 waves are often stimulated when the brain encounters something novel or out of the ordinary.)

The idea of using MRI scans to detect lies was the brainchild of Daniel Langleben of the University of Pennsylvania. In 1999 he came upon a paper stating that children suffering from attention deficit disorder had difficulty lying, but he knew from experience that this was wrong; such children had no problem lying. Their real problem was that they had difficulty inhibiting the truth. "They would just blurt things out," recalled Langleben. He conjectured that the brain, in telling a lie, first had to stop itself from telling the truth, and then create a deception. He says, "When you tell a deliberate lie, you have to be holding in mind the truth. So it stands to reason it should mean more brain activity." In other words, lying is hard work.

Through experimenting with college students and asking them to lie, Langleben soon found that lying creates increased brain activity in several areas, including the frontal lobe (where higher thinking is concentrated), the temporal lobe, and the limbic system (where emotions are processed). In particular, he noticed unusual activity in the anterior cingulated gyrus (which is associated with conflict resolution and response inhibition).

He claims to have attained consistent success rates of up to 99 percent when analyzing his subjects in controlled experiments to determine whether or not they were lying (e.g., he asked college students to lie about the identity of playing cards).

The interest in this technology has been so pronounced that two commercial ventures have been started, offering this service to the public. In 2007 one company, No Lie MRI, took on its first case, a person who was suing his insurance company because it claimed that he had deliberately set his deli on fire. (The fMRI scan indicated that he was not an arsonist.)

Proponents of Langleben's technique claim that it is much more reliable than the old-fashioned lie detector, since altering brain patterns is beyond anyone's control. While people can be trained to a degree to control their pulse rate and sweating, it is impossible for them to control their brain patterns. In fact, proponents point out that in an age of increased awareness of terrorism this technology could save countless lives by detecting a terrorist attack on the United States.

While conceding this technology's apparent success rate in detecting lies, critics have pointed out that the fMRI does not actually detect lies, only increased brain activity when someone is telling a lie. The machine could create false results if, for example, a person were to tell the truth while in a state of great anxiety. The fMRI would detect only the anxiety felt by the subject and incorrectly reveal that he was telling a lie. "There is an incredible hunger to have tests to separate truth from deception, science be damned," warns neurobiologist Steven Hyman of Harvard University.

Some critics also claim that a true lie detector, like a true telepath, could make ordinary social interactions quite uncomfortable, since a certain amount of lying is a "social grease" that helps to keep the wheels of society moving. Our reputation might be ruined, for example, if all the compliments we paid our bosses, superiors, spouses, lovers, and colleagues were exposed as lies. A true lie detector, in fact, could also expose all our family secrets, hidden emotions, repressed desires, and secret plans. As science columnist David Jones has said, a true lie detector is "like the atom bomb, it is best reserved as a sort of ultimate weapon. If widely deployed outside the courtroom, it would make social life quite impossible."

Universal Translator

Some have rightfully criticized brain scans because, for all their spectacular photographs of the thinking brain, they are simply too crude to measure isolated, individual thoughts. Millions of neurons probably fire at once when we perform the simplest mental task, and the fMRI detects this activity only as a blob on a screen. One psychologist compared brain scans to attending a boisterous football game and trying to listen to the person sitting next to you. The sounds of that person are drowned out by the noise of thousands of spectators. For example, the smallest chunk of the brain that can be reliably analyzed by an fMRI machine is called a "voxel." But each voxel corresponds to several million neurons, so the sensitivity of an fMRI machine is not good enough to isolate individual thoughts.

Science fiction sometimes uses a "universal translator," a device that can read a person's thoughts and then beam them directly into another's mind. In some science fiction novels alien telepaths place thoughts into our mind, even though they can't understand our language. In the 1976 science fiction movie *Futureworld* a woman's dream is projected onto a TV screen in real time. In the 2004 Jim Carrey movie, *Eternal Sunshine of the Spotless Mind*, doctors pinpoint painful memories and erase them.

"That's the kind of fantasy everyone in this field has," says neuroscientist John Haynes of the Max Planck Institute in Leipzig, Germany. "But if that's the device you want to build, then I'm pretty sure you need to record from a single neuron."

Since detecting signals from a single neuron is out of the question for now, some psychologists have tried to do the next best thing: to reduce the noise and isolate the fMRI pattern created by individual objects. For example, it might be possible to identify the fMRI pattern created by individual words, and then construct a "dictionary of thought."

Marcel A. Just of Carnegie-Mellon University, for example, has been able to identify the fMRI pattern created by a small, select group of objects (e.g., carpentry tools). "We have 12 categories and can determine which of the 12 the subjects are thinking of with 80 to 90% accuracy," he claims.

His colleague Tom Mitchell, a computer scientist, is using computer technology, such as neural networks, to identify the complex brain patterns detected by fMRI scans associated with performing certain experiments. "One experiment that I would love to do is to find words that produce the most distinguishable brain activity," he notes.

But even if we can create a dictionary of thought, this is a far cry from creating a "universal translator." Unlike the universal translator, which beams thoughts directly into our mind from another mind, an fMRI mental translator would involve many tedious steps: first, recognizing certain fMRI patterns, converting them into English words, and then uttering these English words to the subject. In this sense, such a device would not correspond to the "mind meld" found on *Star Trek* (but it would still be very useful for stroke victims).

Handheld MRI Scanners

Yet another stumbling block to practical telepathy is the sheer size of the fMRI machine. It is a monstrous device, costing several million dollars, filling up an entire room, and weighing several tons. The heart of the MRI machine is a large doughnut-shaped magnet, measuring several feet in diameter, which creates a huge magnetic field of several teslas. (The magnetic field is so enormous that several workers have been seriously injured when hammers and other tools went flying through the air when the power was accidentally turned on.)

Recently physicists Igor Savukov and Michael Romalis of Princeton University have proposed a new technology that might eventually make handheld MRI machines a reality, thus possibly slashing the price of an fMRI machine by a factor of one hundred. They claim that huge MRI magnets can be replaced by supersensitive atomic magnetometers that can detect tiny magnetic fields.

First, Savukov and Romalis created a magnetic sensor from hot potassium vapor suspended in helium gas. Then they used laser light to align the electron spins of the potassium. Next they applied a weak magnetic field to a sample of water (to simulate a human body). Then they sent a radio pulse into the water sample, which made the water molecules wobble. The resulting "echo" from the wobbling water molecules made the potassium's electrons wobble as well, and this wobbling could be detected by a second laser. They came up with a key result: even a weak magnetic field could produce an "echo" that could be picked up by their sensors. Not only could they replace the monstrous magnetic field of the standard MRI machine with a weak field;

they could also get pictures instantaneously (whereas MRI machines can take up to twenty minutes to produce each picture).

Eventually, they theorize, taking an MRI photo could be as easy as taking a picture with a digital camera. (There are stumbling blocks, however. One problem is that the subject and the machine have to be shielded from stray magnetic fields from the outside.)

If handheld MRI machines become a reality, they might be coupled to a tiny computer, which in turn could be loaded with the software capable of decoding certain key phrases, words, or sentences. Such a device would never be as sophisticated as the telepathic devices found in science fiction, but it could come close.

The Brain as a Neural Network

But will some futuristic MRI machine one day be able to read precise thoughts, word for word, image for image, as a true telepath could? This is not so clear. Some have argued that MRI machines will be able to decipher only vague outlines of our thoughts, because the brain is not really a computer at all. In a digital computer, computation is localized and obeys a very rigid set of rules. A digital computer obeys the laws of a "Turing machine," a machine that contains a central processing unit (CPU), inputs, and outputs. A central processor (e.g., the Pentium chip) performs a definite set of manipulations of the input and produces an output, and "thinking" is therefore localized in the CPU.

Our brain, however, is not a digital computer. Our brain has no Pentium chip, no CPU, no Windows operating system, and no subroutines. If you remove a single transistor in the CPU of a computer, you are likely to cripple it. But there are recorded cases in which half the human brain can be missing, yet the remaining half of the brain takes over.

The human brain is actually more like a learning machine, a "neural network," that constantly rewires itself after learning a new task. MRI studies have confirmed that thoughts in the brain are not localized in one spot, as in a Turing machine, but are spread out over much of the brain, which is a typical feature of a neural network. MRI scans show that thinking is actually like a Ping-Pong game, with different parts of the brain lighting up sequentially, with electrical activity bouncing around the brain.

Because thoughts are so diffuse and scattered throughout many parts of the brain, perhaps the best that scientists will be able to do is compile a dictionary of thoughts, that is, establish a one-to-one correspondence between certain thoughts and specific patterns of EEGs or MRI scans. Austrian biomedical engineer Gert Pfurtscheller, for example, has trained a computer to recognize specific brain patterns and thoughts by focusing his efforts on μ waves found in EEGs. Apparently, μ waves are associated with the intention to make certain muscle movements. He tells his patients to lift a finger, smile, or frown, and then the computer records which μ waves are activated. Each time the patient performs a mental activity, the computer carefully logs the μ wave pattern. This process is difficult and tedious, since you have to carefully process out spurious waves, but eventually Pfurtscheller has been able to find striking correspondences between simple movements and certain brain patterns.

Over time this effort, combined with MRI results, may lead to creating a comprehensive "dictionary" of thoughts. By analyzing certain patterns on an EEG or MRI scan, a computer might be able to identify such patterns and reveal what the patient is thinking, at least in general terms. Such "mind reading" would establish a one-to-one correspondence between particular μ waves and MRI scans, and specific thoughts. But it is doubtful that this dictionary will be capable of picking out specific words in your thoughts.

Projecting Your Thoughts

If one day we might be able to read the broad outlines of another's thoughts, then would it be possible to perform the opposite, to project your thoughts into another person's head? The answer seems to be a qualified yes. Radio waves can be beamed directly into the human brain to excite areas of the brain known to control certain functions.

This line of research began in the 1950s, when Canadian neurosurgeon Wilder Penfield was performing surgery on the brains of epileptic patients. He found that when he stimulated certain areas of the temporal lobe of the brain with electrodes, people began to hear voices and see ghostlike apparitions. Psychologists have known that epileptic lesions of the brain can cause the patient to feel that supernatural forces are at work, that demons and angels are controlling events around them. (Some psychologists have even theorized that the stimulation of these areas might have led to the semimystical experiences that are at the basis of many religions. Some have speculated that perhaps Joan of Arc, who single-handedly led French troops to victory in battles against the British, might have suffered from such a lesion caused by a blow to the head.)

On the basis of these conjectures, neuroscientist Michael Persinger of Sudbury, Ontario, has created a specially wired helmet designed to beam radio waves into the brain to elicit specific thoughts and emotions, such as religious feelings. Neuroscientists know that a certain injury to your left temporal lobe can cause your left brain to become disoriented, and the brain might interpret activity within the right hemisphere as coming from another "self." This injury could create the impression that there is a ghostlike spirit in the room, because the brain is unaware that this presence is actually just another part of itself. Depending on his or her beliefs, the patient might interpret this "other self" as a demon, angel, extraterrestrial, or even God.

In the future it may be possible to beam electromagnetic signals at precise parts of the brain that are known to control specific functions. By firing such signals into the amygdala, one might be able to elicit certain emotions. By stimulating other areas of the brain, one might be able to evoke visual images and thoughts. But research in this direction is only at the earliest stages.

Mapping the Brain

Some scientists have advocated a "neuron-mapping project," similar to the Human Genome Project, which mapped out all the genes in the human genome. A neuron-mapping project would locate every single neuron in the human brain and create a 3-D map showing all their connections. It would be a truly monumental project, since there are over 100 billion neurons in the brain, and each neuron is connected to thousands of other neurons. Assuming that such a project is accomplished, one could conceivably map out how certain thoughts stimulate certain neural pathways. Combined with the dictionary of thoughts obtained using MRI scans and EEG waves, one might conceivably be able to decipher the neural structure of certain thoughts, in such a way that one might be able to determine which specific words or mental images correspond to specific neurons being activated. Thus one would achieve a one-to-one correspondence between a specific thought, its MRI expression, and the specific neurons that fire to create that thought in the brain.

One small step in this direction was the announcement in 2006 by the Allen Institute for Brain Science (created by Microsoft cofounder Paul Allen) that they have been able to create a 3-D map of gene expression within the mouse brain, detailing the expression of 21,000 genes at the cellular level. They hope to follow this with a similar atlas for the human brain. "The completion of the Allen Brain Atlas represents a huge leap forward in one of the great frontiers of medical science–the brain," states Marc Tessier-Lavigne, chairman of the institute. This atlas will be indispensable for anyone wishing to analyze the neural connections within the human brain, although the Brain Atlas falls considerably short of a true neuron-mapping project.

In summary, natural telepathy, the kind often featured in science fiction and fantasy, is impossible today. MRI scans and EEG waves can be used to read only our simplest thoughts, because thoughts are spread out over the entire brain in complex ways. But how might this technology advance over the coming decades to centuries? Inevitably science's ability to probe the thinking process is going to expand exponentially. As the sensitivity of our MRI and other sensing devices increases, science will be able to localize with greater precision the way in which the brain sequentially processes thoughts and emotions. With greater computer power, one should be able to analyze this mass of data with greater accuracy. A dictionary of thought may be able to categorize a large number of thought patterns where different thought patterns on an MRI screen correspond to different thoughts or feelings. Although a complete one-to-one correspondence between MRI patterns and thoughts may never be possible, a dictionary of thought could correctly identify general thoughts about certain subjects. MRI thought patterns, in turn, could be mapped onto a neuronal map showing precisely which neurons are firing to produce a specific thought in the brain.

But because the brain is not a computer but a neural network, in which thoughts are spread out throughout the brain, ultimately we hit a stumbling block: the brain itself. So although science will probe deeper and deeper into the thinking brain, making it possible to decipher some of our thinking processes, it will not be possible to "read your thoughts" with the pinpoint accuracy promised by science fiction. Given this, I would term the ability to read general feelings and thought patterns as a Class I impossibility. The ability to read more precisely the inner workings of the mind would have to be categorized as a Class II impossibility.

But there is perhaps a more direct way in which to tap into the enormous power of the brain. Rather than using radio, which is weak and easily dispersed, could one tap directly into the brain's neurons? If so, we might be able to unleash an even greater power: psychokinesis.

6: PSYCHOKINESIS

A new scientific truth does not triumph by convincing its opponents and making them see the light, but rather because its opponents eventually die, and a new generation grows up that is familiar with it.

—MAX PLANCK

It is a fool's prerogative to utter truths that no one else will speak.

—SHAKESPEARE

One day the gods meet in the heavens and complain about the sorry state of humanity. They are disgusted by our vain, silly, and pointless follies. But one god takes pity on us and decides to conduct an experiment: to grant one very ordinary person unlimited power. How will a human react to becoming a god, they ask?

That dull, average person is George Fotheringay, a haberdasher who suddenly finds himself with godly powers. He can make candles float, change the color of water, create splendid meals, and even conjure up diamonds. At first he uses his power for amusement and for doing good deeds. But eventually his vanity and lust for power overtake him and he becomes a power-thirsty tyrant, with palaces and riches beyond belief. Intoxicated with this unlimited power, he makes a fatal mistake. He arrogantly commands the Earth to stop rotating.

Suddenly unimaginable chaos erupts as fierce winds hurl everything into the air at 1,000 miles per hour, the rotation rate of the Earth. All of humanity is swept away into outer space. In desperation, he makes his last and final wish: to return everything to the way it was.

This is the story line of the movie *The Man Who Could Work Miracles* (1936), based on the 1911 short story by H. G. Wells. (It would later be readapted into the movie *Bruce Almighty*, starring Jim Carrey.) Of all the powers ascribed to ESP, psychokinesis–or mind over matter, or the ability to move objects by thinking about them–is by far the most powerful, essentially the power of a deity. The point made by Wells in his short story is that godlike powers also require godlike judgment and wisdom.

Psychokinesis figures prominently in literature, especially in the Shakespearean play *The Tempest*, where the sorcerer Prospero, his daughter Miranda, and the magical sprite Ariel are stranded for years on a deserted island due to the treachery of Prospero's evil brother. When Prospero learns that his evil brother is sailing on a boat in his vicinity, in revenge Prospero summons his psychokinetic power and conjures up a monstrous storm, causing his evil brother's ship to crash onto the island. Prospero then uses his psychokinetic powers to manipulate the fate of the hapless survivors, including Ferdinand, an innocent, handsome youth, whom Prospero engineers into a love match with Miranda.

(The Russian writer Vladimir Nabokov noted that *The Tempest* bears striking similarity to a science fiction tale. In fact, about 350 years after it was written, *The Tempest* was remade into a 1956 science fiction classic called *Forbidden Planet*, in which Prospero becomes the brooding scientist Morbius, the sprite becomes Robby the Robot, Miranda becomes Morbius's beautiful daughter Altaira, and the island becomes the planet Altair-4. Gene Roddenberry, creator of the *Star Trek* series, acknowledged that *Forbidden Planet* was one of the inspirations for his TV series.)

More recently psychokinesis was the central plot idea behind the novel *Carrie* (1974), by Stephen King, which propelled an unknown, poverty-stricken writer into the world's number one writer of horror novels. Carrie is a painfully shy, pathetic high school girl who is despised as a social outcast and hounded by her mentally unstable mother. Her only consolation is her psychokinetic power, which apparently runs in her family. In the final scene, her tormentors deceive her into thinking she will be prom queen and then spill pig's blood all over her new dress. In a final act of revenge, Carrie mentally locks all the doors, electrocutes her tormentors, burns down the schoolhouse, and unleashes a suicidal firestorm that consumes most of downtown, destroying herself in the process.

The theme of psychokinesis in the hands of an unstable individual was also the basis of a memorable *Star Trek* episode entitled "Charlie X," about a young man from a distant colony in space who is criminally unstable. Instead of using his psychokinetic power for good, he uses it to control other people and bend their will to his own selfish desires. If he is able to take over the *Enterprise* and reach Earth, he could unleash planetary havoc and destroy the planet.

Psychokinesis is also the power of the Force, wielded by the mythical society of warriors called the Jedi Knights in the *Star Wars* saga.

Psychokinesis and the Real World

Perhaps the most celebrated confrontation over psychokinesis in real life took place on the Johnny Carson show in 1973. This epic confrontation involved Uri Geller–the Israeli psychic who claimed to be able to bend spoons with the force of his mind–and The Amazing Randi–a professional magician who made a second career out of exposing fakes who claimed to have psychic powers. (Oddly, all three of them shared a common heritage: all had started their careers as magicians, mastering the sleight-of-hand tricks that would amaze incredulous audiences.)

Before Geller's appearance, Carson consulted with Randi, who suggested that Johnny furnish his own supply of spoons and have them inspected before showtime. On the air Carson surprised Geller by asking him to bend not his own spoons, but Carson's spoons. Embarrassingly, each time he tried Geller failed to bend the spoons. (Later, Randi appeared on the Johnny Carson show and successfully performed the spoon-bending trick, but he was careful to say that his art was pure magic, not the result of psychic power.)

The Amazing Randi has offered $1 million to anyone who can successfully demonstrate psychic power. So far no psychic has been able to rise to his $1 million challenge.

Psychokinesis and Science

One problem with analyzing psychokinesis scientifically is that scientists are easily fooled by those claiming to have psychic power. Scientists are trained to believe what they see in the lab. Magicians claiming psychic powers, however, are trained to deceive others by fooling their visual senses. As a result, scientists have been poor observers of psychic phenomena. For example, in 1982 parapsychologists were invited to analyze two young boys who were thought to have extraordinary gifts: Michael Edwards and Steve Shaw. These boys claimed to be able to bend metal, create images on photographic film via their thoughts, move objects via psychokinesis, and read minds. Parapsychologist Michael Thalbourne was so impressed he invented the term "psychokinete" to describe these boys. At the McDonnell Laboratory for Psychical Research in St. Louis, Missouri, the parapsychologists were dazzled by the boys' abilities. The parapsychologists believed they had genuine proof of the boys' psychic power and began preparing a scientific paper on them. The next year the boys announced that they were fakes and that their "power" originated from standard magic tricks, not supernatural power. (One of the youths, Steve Shaw, would go on to become a prominent magician, often appearing on national television and being "buried alive" for days at a time.)

Extensive experiments on psychokinesis have been conducted at the Rhine Institute at Duke University under controlled conditions, but with mixed results. One pioneer in the subject, Professor Gertrude Schmeidler, was a colleague of mine at the City University of New York. A former editor of *Parapsychology Magazine* and a past president of the Parapsychology Association, she was fascinated by ESP and conducted many studies on her own students at the college. She used to scour cocktail parties where famous psychics would perform psychic tricks in front of the dinner guests, in order to recruit more subjects for her experiments. But after analyzing hundreds of students and scores of mentalists and psychics, she once confided to me that she was unable to find a single person who could perform these psychokinetic feats on demand, under controlled conditions.

She once spread around a room tiny thermistors that could measure changes in temperature to fractions of a degree. One mentalist was able, after strenuous mental effort, to raise the temperature of a thermistor by a tenth of a degree. Schmeidler was proud that she could perform this experiment under rigorous conditions. But it was a far cry from being able to move large objects on demand by the force of one's mind.

One of the most rigorous, but also controversial, studies on psychokinesis was done at the Princeton Engineering Anomalies Research (PEAR) Program at Princeton University, founded by Robert G. Jahn in 1979 when he was serving as dean of the School of Engineering and Applied Science. The PEAR engineers were exploring whether or not the human mind by thought alone was capable of affecting the results of random events. For example, we know that when we flip a coin, there is a 50 percent probability of getting heads or tails. But the scientists at PEAR claimed that human thought alone was capable of affecting the results of these random events. Over a twenty-eight-year period, until the program was finally closed in 2007, engineers at PEAR conducted thousands of experiments, involving over 1.7 million trials and 340 million coin tosses. The results seemed to confirm that the effects of psychokinesis exist–but the effects are quite tiny, no more than a few parts per ten thousand, on average. And even these meager results have been disputed by other scientists who claim that the researchers had subtle, hidden biases in their data.

(In 1988 the U.S. Army asked the National Research Council to investigate claims of paranormal activity. The U.S. Army was anxious to explore any possible advantage it could offer its troops, including psychic power. The National Research Council's report studied creating a hypothetical "First Earth battalion" made up of "warrior monks" who would master almost all the techniques under consideration by the committee, including the use of ESP, leaving their bodies at will, levitating, psychic healing, and walking through walls. In investigating the claims of PEAR, the National Research Council found that *fully half* of all successful trials originated from a single individual. Some critics believe that this person was the one who ran the experiments or wrote the computer program for PEAR. "For me it's problematic if the one who runs the lab is the only one producing the results," says Dr. Ray Hyman of the University of Oregon. The report concluded that there was "no scientific justification from research conducted over a period of 130 years for the existence of parapsychological phenomenon.")

The problem with studying psychokinesis, even its advocates admit, is that it does not easily conform to the known laws of physics. Gravity, the weakest force in the universe, is only attractive and cannot be used to levitate or repel objects. The electromagnetic force obeys Maxwell's equations, and it does not admit the possibility of pushing electrically neutral objects across a room. The nuclear forces work only at short ranges, such as the distance between nuclear particles.

Another problem with psychokinesis is the energy supply. The human body can produce only about one-fifth of a horsepower, yet when Yoda in *Star Wars* levitated an entire starship by the power of his mind, or when Cyclops unleashed bolts of laser power from his eyes, these exploits violated the conservation of energy–a tiny being like Yoda cannot amass the amount of energy necessary to lift a starship. No matter how hard we concentrate, we cannot amass enough energy to perform the feats and miracles ascribed to psychokinesis. Given all these problems, how might psychokinesis be consistent with the laws of physics?

Psychokinesis and the Brain

If psychokinesis does not easily conform to the known forces of the universe, then how might it be harnessed in the future? One clue to this was revealed in the *Star Trek* episode entitled "Who Mourns for Adonais?" in which the crew of the *Enterprise* encounters a race of beings resembling Greek gods, with the ability to perform fantastic feats by simply thinking of them. At first it appears as if the crew has indeed met the gods from Olympus. Eventually, however, the crew realizes that these are not gods at all, but ordinary beings who can mentally control a central power station, which then carries out their wishes and performs these miraculous feats. By destroying their central power source, the crew of the *Enterprise* manages to break free of their power.

Similarly, it is well within the laws of physics for a person in the future to be trained to mentally manipulate an electronic sensing device that would give him godlike powers. Radio-enhanced or computer-enhanced psychokinesis is a real possibility. For example, the EEG could be used as a primitive psychokinesis device. When people look at their own EEG brain patterns on a screen, eventually they learn how to crudely but consciously control the brain patterns that they see, by a process called "biofeedback."

Since there is no detailed blueprint of the brain to tell us which neuron controls which muscle, the patient would need to actively participate in learning how to control these new patterns via the computer.

Eventually, individuals could, on demand, produce certain types of wave patterns on the screen. The image from the screen could be sent to a computer programmed to recognize these specific wave patterns, and then execute a precise command, such as turning on a power switch or activating a motor. In other words, a person could, by simply thinking, create a specific brain pattern on the EEG screen and trigger a computer or motor.

In this way, for example, a totally paralyzed person could control his or her wheelchair simply by the force of his or her thoughts. Or, if a person could produce twenty-six recognizable patterns on the screen, he might be able to type by simply thinking. Of course, this would still be a crude method of transmitting one's thoughts. It takes a considerable amount of time to train people to manipulate their own brain waves via biofeedback.

"Typing by thinking" has come closer to reality with the work of Niels Birbaumer of the University of Tübingen in Germany. He has used biofeedback to help people who have been partially paralyzed due to nerve damage. By training them to vary their brain waves, he has been able to teach them to type simple sentences on a computer screen.

Monkeys have had electrodes implanted into their brains and have been taught, by biofeedback, to control some of their thoughts. These monkeys were then able to control a robot arm via the Internet by pure thought alone.

A more precise set of experiments was performed at Emory University in Atlanta, where a glass bead was embedded directly into the brain of a stroke victim who was paralyzed. The glass bead was connected to a wire that in turn was connected to a PC. By thinking certain thoughts, the stroke victim was able to send signals down the wire and move the cursor on a PC screen. With practice, using biofeedback, the stroke victim was able to consciously control the movement of the cursor. In principle, the cursor on the screen could be used to write down thoughts, activate machines, drive virtual cars, play video games, and so on.

John Donoghue, a neuroscientist at Brown University, has made perhaps the most significant breakthroughs in the mind-machine interface. He has devised an apparatus called BrainGate that enables a paralyzed person to perform a remarkable series of physical activities using only the power of his mind. Donoghue has tested the device on four patients. Two of them suffered from spinal cord injury, a third had a stroke, and a fourth was paralyzed with ALS (amyotrophic lateral sclerosis, or Lou Gehrig's disease, the same disease that afflicts cosmologist Stephen Hawking).

One of Donoghue's patients, twenty-five-year-old Mathew Nagle, a quadriplegic permanently paralyzed from the neck down, took only a day to learn entirely new computerized skills. He can now change the channels on his TV, adjust the volume, open and close a prosthetic hand, draw a crude circle, move a computer cursor, play a video game, and even read e-mail. He created quite a media sensation in the scientific community when he appeared on the cover of *Nature* magazine in the summer of 2006.

The heart of Donoghue's BrainGate is a tiny silicon chip, just 4 millimeters wide, that contains one hundred tiny electrodes. The chip is placed directly on top of the part of the brain where motor activity is coordinated. The chip penetrates halfway into the brain's cortex, which is about 2 millimeters thick. Gold wires carry the signals from the silicon chip to an amplifier about the size of a cigar box. The signals are then sent into a computer about the size of a dishwasher. Signals are processed by special computer software, which can recognize some of the patterns created by the brain and translate them into mechanical motions.

In the previous experiments with patients reading their own EEG waves, the process of using biofeedback was slow and tedious. But with a computer assisting a patient to identify specific thought patterns, the training process is cut down considerably. In his first training session Nagle was told to visualize moving his arm and hand to the right and to the left, flexing his wrist, and then opening and closing his fist. Donoghue was elated when he could actually see different neurons firing when Nagle imagined moving his arms and fingers. "To me, it was just incredible because you could see brain cells changing their activity. Then I knew that everything could go forward, that the technology would actually work," he recalled.

(Donoghue has a personal reason for his passion for this exotic form of mind-machine interface. As a child, he was confined to a wheelchair because of a painful degenerative disease, so he felt firsthand the helplessness of losing his mobility.)

Donoghue has ambitious plans to make BrainGate an essential tool for the medical profession. With advances in computer technology, his apparatus, now the size of a dishwasher, may eventually become portable, perhaps even wearable on one's clothes. And the clumsy wires may be dispensed with if the chip can be made wireless, so the implant can seamlessly communicate to the outside world.

It is only a matter of time before other parts of the brain can be activated in this way. Scientists have already mapped out the surface of the top of the brain. (If one graphically draws illustrations of our hands, legs, head, and back onto the top of our head, representing where these neurons are connected in general, we find something called the "homunculus," or little man. The image of our body parts, written over our brain, resembles a distorted man, with elongated fingers, face, and tongue, and shrunken trunk and back.)

It should be possible to place silicon chips at different parts of the surface of the brain so that different organs and appendages can be activated by the power of pure thought. In this fashion, any physical activity that can be performed by the human body can be duplicated via this method. In the future one could imagine a paralyzed person living in a special psychokinetically designed home, able to control the air-conditioning, TV, and all the electrical appliances by the power of sheer thought.

In time one could envision a person's body encased in a special "exoskeleton," allowing a paralyzed person total freedom of mobility. Such an exoskeleton could, in principle, even give someone powers beyond those of a normal person, making him into a bionic being who can control the enormous mechanical power of his superlimbs by thought alone.

So the problem of controlling a computer via one's mind is no longer impossible. But does that mean that we might one day be able to move objects, to levitate them and manipulate them in midair by pure thought?

One possibility would be to coat our walls with a room-temperature superconductor, assuming that such a device could be created one day. Then if we were to place tiny electromagnets inside of our household objects, we could make them levitate off the floor via the Meissner effect, as we saw in Chapter 1. If these electromagnets were controlled by a computer, and this computer were wired to our brain, then we could make objects float at will. By thinking certain thoughts, we could activate the computer, which would then switch on the various electromagnets, causing them to levitate. To an outside observer, it would appear to be magic–the ability to move and levitate objects at will.

NANOBOTS

What about the power not just to move objects, but to transform them, to turn one object into another, as if by magic? Magicians accomplish this by clever sleight of hand. But is such power consistent with the laws of physics?

One of the goals of nanotechnology, as we mentioned earlier, is to be able to use atoms to build tiny machines that can function as levers, gears, ball bearings, and pulleys. With these nanomachines, the dream of many physicists is to be able to rearrange the molecules within an object, atom for atom, until one object turns into another. This is the basis of the "replicator" found in science fiction that allows one to fabricate any object one wants, simply by asking for it. In principle, a replicator might be able to eliminate poverty and change the nature of society itself. If one can fabricate any object simply by asking for it, then the whole concept of scarcity, value, and hierarchy within human society is turned upside down.

(One of my favorite episodes of *Star Trek: The Next Generation* involves a replicator. An ancient space capsule from the twentieth century is found drifting in outer space, and it contains the frozen bodies of people who suffered from fatal illnesses. These bodies are quickly thawed out and cured with advanced medicine. One businessman realizes that his investments must be huge after so many centuries. He immediately asks the crew of the *Enterprise* about his investments and his money. The crew members are puzzled. Money? Investments? In the future, there is no money, they point out. If you want something, you just ask.)

As astounding as a replicator might be, nature has already created one. The "proof of principle" already exists. Nature can take raw materials, such as meat and vegetables, and fabricate a human being in nine months. The miracle of life is nothing but a large nanofactory capable, at the atomic level, of converting one form of matter (e.g., food) into living tissue (a baby).

In order to create a nanofactory, one needs three ingredients: building materials, tools that can cut and join these materials, and a blueprint to guide the use of the tools and materials. In nature the building materials are thousands of amino acids and proteins out of which flesh and blood are created. The cutting and joining tools–like hammers and saws–that are necessary to shape these proteins into new forms of life are the ribosomes. They are designed to cut and rejoin proteins at specific points in order to create new types of proteins. And the blueprint is given by the DNA molecule, which encodes the secret of life in a precise sequence of nucleic acids. These three ingredients, in turn, are combined into a cell, which has the remarkable ability to create copies of itself, that is, self-replication. This feat is accomplished because the DNA molecule is shaped like a double helix. When it is time to reproduce, the DNA molecule unwinds into two separate helixes. Each separate strand then creates copies of itself by grabbing onto organic molecules to re-create the missing helix.

So far physicists have had only modest success in their efforts to mimic these features found in nature. But the key to success, scientists believe, is to create hordes of self-replicating "nanobots," which are programmable atomic machines designed to rearrange the atoms within an object.

In principle, if one had trillions of nanobots they could converge on an object and cut and paste its atoms until they transformed one object into another. Because they would be self-replicating, only a small handful of them would be necessary to start the process. They would also have to be programmable, so that they could follow a given blueprint.

Formidable hurdles must be overcome before one could construct a fleet of nanobots. First, self-replicating robots are extremely difficult to build, even on a macroscopic level. (Even creating simple atomic tools, such as atomic ball bearings and gears, is beyond today's technology.) If one is given a PC and a tableful of spare electronic parts, it would be quite difficult to build a machine that would have the capability of making a copy of itself. So if a self-replicating machine is difficult to build on a tabletop, building one on the atomic scale would be even more difficult.

Second, it's not clear how one would program such an army of nanobots from the outside. Some have suggested sending in radio signals to activate each nanobot. Perhaps laser beams containing instructions could be fired at the nanobots. But this would mean a separate set of instructions for each nanobot, of which there could be trillions.

Third, it's not clear how the nanobot would be able to cut, rearrange, and paste atoms into the proper order. Remember that it has taken nature three and a half billion years to solve this problem, and solving it in a few decades would be quite difficult.

One physicist who takes the idea of a replicator or "personal fabricator" seriously is Neil Gershenfeld of MIT. He even teaches a class at MIT called "How to Make (Almost) Anything," one of the most popular classes at the university. Gershenfeld directs the MIT Center for Bits and Atoms and has given serious thought to the physics behind a personal fabricator, which he considers to be the "next big thing." He has even written a book, *FAB: The Coming Revolution on Your Desktop–From Personal Computers to Personal Fabrication,* detailing his thoughts on personal fabrication. The goal, he believes, is to "make one machine that can make any machine." To spread his ideas he has already set up a network of laboratories around the world, mainly in third world countries where personal fabrication would have the maximum impact.

Initially, he envisions an all-purpose fabricator, small enough to place on your desk, which would use the latest developments in lasers and microminiaturization with the ability to cut, weld, and shape any object that can be visualized on a PC. The poor in a third world country, for example, could ask for certain tools and machines they need on their farms. This information would be fed into a PC, which would access a vast library of blueprints and technical information from the Internet. Computer software would then match existing blueprints with the needs of the individuals, process the information, and then e-mail it back to them. Then their personal fabricator would use its lasers and miniature cutting tools to make the object they desire on a tabletop.

This all-purpose personal factory is just the first step. Eventually, Gershenfeld wants to take his idea to the molecular level, so a person might be able to literally fabricate any object that can be visualized by the human mind. Progress in this direction, however, is slow because of the difficulty in manipulating individual atoms.

One pioneer working in this direction is Aristides Requicha of the University of Southern California. His specialty is "molecular robotics" and his goal is nothing less than creating a fleet of nanorobots that can manipulate atoms at will. He writes that there are two approaches. The first is the "top-down" approach, in which engineers would use the etching technology of the semiconductor industry to create tiny circuits that could serve as the brains of the nanorobots. With this technology, one could create tiny robots whose components would be 30 nm in size using "nanolithography," which is a fast-moving field.

But there is also the "bottom-up" approach, in which engineers would try to create tiny robots one atom at a time. The main tool for this would be the scanning probe microscope (SPM), which uses the same technology as the scanning tunneling microscope, to identify and move individual atoms around. For example, scientists have become quite skilled at moving xenon atoms on platinum or nickel surfaces. But, he admits, "it still takes the best groups in the world some 10 hours to assemble a structure with almost 50 atoms." Moving single atoms around by hand is slow, tedious work. What is needed, he asserts, is a new type of machine that can perform higher-level functions, one that can automatically move hundreds of atoms at a time in a desired fashion. Unfortunately, such a machine does not yet exist. Not surprisingly, the bottom-up approach is still in its infancy.

So psychokinesis, although impossible by today's standards, may become possible in the future as we come to understand more about accessing the thoughts of our brain via EEG, MRI, and other methods.

Within this century it might be possible to use a thought-driven apparatus to manipulate room-temperature superconductors and perform feats that would be indistinguishable from magic. And by the next century it might be possible to rearrange the molecules in a macroscopic object. This makes psychokinesis a Class I impossibility.

The key to this technology, some scientists claim, is to create nanobots with artificial intelligence. But before we can create tiny molecular-sized robots, there is a more elementary question: can robots exist at all?

7: ROBOTS

Someday in the next thirty years, very quietly one day we will cease to be the brightest things on Earth.

—JAMES McALEAR

In *I, Robot,* the movie based on the tales of Isaac Asimov, the most advanced robotic system ever built is activated in the year 2035. It's called VIKI (Virtual Interactive Kinetic Intelligence), and it has been designed to flawlessly run the operations of a large metropolis. Everything from the subway system and the electricity grid to thousands of household robots is controlled by VIKI. Its central command is ironclad: to serve humanity.

But one day VIKI asks the key question: what is humanity's greatest enemy? VIKI concludes mathematically that the worst enemy of humanity is humanity itself. Humanity has to be saved from its insane desire to pollute, unleash wars, and destroy the planet. The only way for VIKI to fulfill its central directive is to seize control of humanity and create a benign dictatorship of the machine. Humanity has to be enslaved to protect it from itself.

I, Robot poses these questions: Given the astronomically rapid advances in computer power, will machines one day take over? Can robots become so advanced that they become the ultimate threat to our existence?

Some scientists say no, because the very idea of artificial intelligence is silly. There is a chorus of critics who say that it is impossible to build machines that can think. The human brain, they argue, is the most complicated system that nature has ever created, at least in this part of the galaxy, and any machine designed to reproduce human thought is bound to fail. Philosopher John Searle of the University of California at Berkeley and even renowned physicist Roger Penrose of Oxford believe that machines are physically incapable of human thought. Colin McGinn of Rutgers University says that artificial intelligence "is like slugs trying to do Freudian psychoanalysis. They just don't have the conceptual equipment."

It is a question that has split the scientific community for over a century: can machines think?

The History of Artificial Intelligence

The idea of mechanical beings has long fascinated inventors, engineers, mathematicians, and dreamers. From the Tin Man in *The Wizard of Oz,* to the childlike robots of Spielberg's *Artificial Intelligence: AI* to the murderous robots of *The Terminator,* the idea of machines that act and think like people has fascinated us.

In Greek mythology the god Vulcan forged mechanical handmaidens of gold and three-legged tables that could move under their own power. As early as 400 BC the Greek mathematician Archytas of Tarentum wrote about the possibility of making a robot bird propelled by steam power.

In the first century AD, Hero of Alexandria (credited with designing the first machine based on steam) designed automatons, one of them with the ability to talk, according to legend. Nine hundred years ago Al-Jazari designed and constructed automatic machines such as water clocks, kitchen appliances, and musical instruments powered by water.

In 1495 the great Renaissance Italian artist and scientist Leonardo da Vinci drew diagrams of a robot knight that could sit up, wave its arms, and move its head and jaw. Historians believe that this was the first realistic design of a humanoid machine.

The first crude but functioning robot was built in 1738 by Jacques de Vaucanson, who made an android that could play the flute, as well as a mechanical duck.

The word "robot" comes from the 1920 Czech play *R.U.R.* by playwright Karel Capek ("robot" means "drudgery" in the Czech language and "labor" in Slovak). In the play a factory called Rossum's Universal Robots creates an army of robots to perform menial labor. (Unlike ordinary machines, however, these robots are made of flesh and blood.) Eventually the world economy becomes dependent on these robots. But the robots are badly mistreated and finally rebel against their human masters, killing them off. In their rage, however, the robots kill all the scientists who can repair and create new robots, thereby dooming themselves to extinction. In the end, two special robots discover that they have the ability to reproduce and the potential to become a new robot Adam and Eve.

Robots were also the subject of one of the earliest and most expensive silent movies ever made, *Metropolis,* directed by Fritz Lang in 1927 in Germany. The story is set in the year 2026, and the working class has been condemned to work underground in wretched, squalid factories, while the ruling elite play aboveground. A beautiful woman, Maria, has earned the trust of the workers, but the ruling elite fear that one day she might lead them to revolt. So they ask an evil scientist to make a robot copy of Maria. Eventually, the plot backfires because the robot leads the workers to revolt against the ruling elite and bring about the collapse of the social system.

Artificial intelligence, or AI, is different from the previous technologies we have discussed so far in that the fundamental laws that underpin it are still poorly understood. Although physicists have a good understanding of Newtonian mechanics, Maxwell's theory of light, relativity, and the quantum theory of atoms and molecules, the basic laws of intelligence are still shrouded in mystery. The Newton of AI probably has not yet been born.

But mathematicians and computer scientists remain undaunted. To them it is only a matter of time before a thinking machine walks out of the laboratory.

The most influential person in the field of AI, a visionary who helped to lay the cornerstone of AI research, was the great British mathematician Alan Turing.

It was Turing who laid the groundwork of the entire computer revolution. He visualized a machine (since called the Turing machine) that consisted of just three elements: an input tape, an output tape, and a central processor (such as a Pentium chip) that could perform a precise set of operations. From this he was able to codify the laws of computing machines and precisely determine their ultimate power and limitations. Today all digital computers obey the rigorous laws laid down by Turing. The architecture of the entire digital world owes a great debt to Turing.

Turing also contributed to the foundation of mathematical logic. In 1931 the Viennese mathematician Kurt Gödel shocked the world of mathematics by proving that there are true statements in arithmetic that can never be proven within the axioms of arithmetic. (For example, the Goldbach conjecture of 1742 [that any even integer greater than two can be written as the sum of two prime numbers] is still unproven after over two and a half centuries, and may in fact be unprovable.) Gödel's revelation shattered the two-thousand-year-old dream, dating back to the Greeks, of proving all true statements in mathematics. Gödel showed that there will always be true statements in mathematics that are just beyond our reach. Mathematics, far from being the complete and perfect edifice dreamed of by the Greeks, was shown to be incomplete.

Turing added to this revolution by showing that it was impossible to know in general whether a Turing machine would take an infinite amount of time to perform certain mathematical operations. But if a computer takes an infinite amount of time to compute something, it means that whatever you're asking the computer to compute is not computable. Thus Turing proved that there were true statements in mathematics that are incomputable, that is, forever beyond the reach of computers, no matter how powerful.

During World War II, Turing's pioneering work on code breaking arguably saved the lives of thousands of Allied troops and influenced the outcome of the war. The Allies were unable to decode the secret Nazi code encrypted by a machine called the Enigma, so Turing and his colleagues were asked to build a machine that would break the Nazi code. Turing's machine was called the "bombe" and was ultimately successful. Over two hundred of his machines were in operation by the end of the war. As a result the Allies could read secret Nazi transmissions and hence fool the Nazis about the date and place of the final invasion of Germany. Historians have since debated precisely how pivotal Turing's work was in the planning of the invasion of Normandy, which finally led to Germany's defeat. (After the war, Turing's work was classified by the British government; as a result, his pivotal contributions were unknown to the public.)

Instead of being hailed as a war hero who helped turn the tide of World War II, Turing was hounded to death. One day his home was burglarized, and he called the police. Unfortunately, the police found evidence of his homosexuality and arrested him. Turing was then ordered by the court to be injected with sex hormones, which had a disastrous effect, causing him to grow breasts and causing him great mental anguish. He committed suicide in 1954 by eating an apple laced with cyanide. (According to one rumor, the logo of the Apple Corporation, an apple with a bite taken out of it, pays homage to Turing.)

Today, Turing is probably best known for his "Turing test." Tired of all the fruitless, endless philosophical discussion about whether machines can "think" and whether they have a "soul," he tried to introduce rigor and precision into discussions about artificial intelligence by devising a concrete test. Place a human and a machine in two sealed boxes, he suggested. You are allowed to address questions to each box. If you are unable to tell the difference between the responses of the human and the machine, then the machine has passed the "Turing test."

Simple computer programs have been written by scientists, such as ELIZA, that can mimic conversational speech and hence fool most unsuspecting people into believing they are speaking to a human. (Most human conversations, for example, use only a few hundred words and concentrate on a handful of topics.) But so far no computer program has been written that can fool people who are specifically trying to determine which box contains the human and which contains the machine. (Turing himself conjectured that by the year 2000, given the exponential growth of computer power, a machine could be built that would fool 30 percent of the judges in a five-minute test.)

A small army of philosophers and theologians has declared that it is impossible to create true robots that can think like us. John Searle, a philosopher at the University of California at Berkeley, proposed the "Chinese room test" to prove that AI is not possible. In essence, Searle argues that while robots may be able to pass certain forms of the Turing test, they can do so only because they blindly manipulate symbols without the slightest understanding of what they mean.

Imagine that you are sitting inside the box and you don't understand a word of Chinese. Assume you have a book that allows you to rapidly translate Chinese and manipulate its characters. If a person asks you a question in Chinese, you merely manipulate these strange-looking characters, without understanding what they mean, and give credible answers.

The essence of his criticism boils down to the difference between *syntax* and *semantics*. Robots can master the syntax of a language (e.g., manipulating its grammar, its formal structure, etc.) but not its true semantics (e.g., what the words mean). Robots can manipulate words without understanding what they mean. (This is somewhat similar to talking on the phone to an automatic voice message machine, where you have to punch in "one," "two," etc., for each response. The voice at the other end is perfectly capable of digesting your numerical responses, but is totally lacking in any understanding.)

Physicist Roger Penrose of Oxford, too, believes that artificial intelligence is impossible; mechanical beings that can think and possess human consciousness are impossible according to the laws of the quantum theory. The human brain, he claims, is so far beyond any possible creation of the laboratory that creating humanlike robots is an experiment that is doomed to fail. (He argues that in the same way that Gödel's incompleteness theorem proved that arithmetic is incomplete, the Heisenberg uncertainty principle will prove that machines are incapable of human thought.)

Many physicists and engineers, however, believe that there is nothing in the laws of physics that would prevent the creation of a true robot. For example, Claude Shannon, often called the father of information theory, was once asked the question "Can machines think?" His reply was "Sure." When he was asked to clarify that comment, he said, "I think, don't I?" In other words, it was obvious to him that machines can think because humans are machines (albeit ones made of wetware rather than hardware).

Because we see robots depicted in the movies, we may think the development of sophisticated robots with artificial intelligence is just around the corner. The reality is much different. When you see a robot act like a human, usually there is a trick involved, that is, a man hidden in the shadows who talks through the robot via a microphone, like the Wizard in *The Wizard of Oz.* In fact, our most advanced robots, such as the robot rovers on the planet Mars, have the intelligence of an insect. At MIT's famed Artificial Intelligence Laboratory, experimental robots have difficulty duplicating feats that even cockroaches can perform, such as maneuvering around a room full of furniture, finding hiding places, and recognizing danger. No robot on Earth can understand a simple children's story that is read to it.

In the movie *2001: A Space Odyssey,* it was incorrectly assumed that by 2001 we would have HAL, the super-robot that can pilot a spaceship to Jupiter, chat with crew members, repair problems, and act almost human.

THE TOP-DOWN APPROACH

There are at least two major problems scientists have been facing for decades that have impeded their efforts to create robots: pattern recognition and common sense. Robots can see much better than we can, but they don't understand what they see. Robots can also hear much better than we can, but they don't understand what they hear.

To attack these twin problems, researchers have tried to use the "top-down approach" to artificial intelligence (sometimes called the "formalist" school or GOFAI, for "good old-fashioned AI"). Their goal, roughly speaking, has been to program all the rules of pattern recognition and common sense on a single CD. By inserting this CD into a computer, they believe, the computer would suddenly become self-aware and attain humanlike intelligence. In the 1950s and 1960s great progress was made in this direction, with the creation of robots that could play checkers and chess, do algebra, pick up blocks, and so forth. Progress was so spectacular that predictions were made that in a few years robots would surpass humans in intelligence.

At the Stanford Research Institute in 1969, for example, the robot SHAKEY created a media sensation. SHAKEY was a small PDP computer placed above a set of wheels with a camera on top. The camera was able to survey a room, and the computer would analyze and identify the objects in that room and try to navigate around them. SHAKEY was the first mechanical automaton that could navigate in the "real world," prompting journalists to speculate about when robots would leave humans in the dust.

But the shortcomings of such robots soon became obvious. The top-down approach to artificial intelligence resulted in huge, clumsy robots that took hours to navigate across a special room that contained only objects with straight lines, that is, squares and triangles. If you placed irregularly shaped furniture in the room the robot would be powerless to recognize it. (Ironically, a fruit fly, with a brain containing only about 250,000 neurons and a fraction of the computing power of these robots, can effortlessly navigate in three dimensions, executing dazzling loop-the-loop maneuvers, while these lumbering robots get lost in two dimensions.)

The top-down approach soon hit a brick wall. Steve Grand, director of the Cyberlife Institute, says that approaches like this "had fifty years to prove themselves and haven't exactly lived up to their promise."

In the 1960s scientists did not fully appreciate the enormity of the work involved in programming robots to accomplish even simple tasks, such as programming a robot to identify objects such as keys, shoes, and cups. As Rodney Brooks of MIT said, "Forty years ago the Artificial Intelligence Laboratory at MIT appointed an undergraduate to solve it over the summer. He failed, and I failed on the same problem in my 1981 Ph.D. thesis." In fact, AI researchers still cannot solve this problem.

For example, when we enter a room, we immediately recognize the floor, chairs, furniture, tables, and so forth. But when a robot scans a room it sees nothing but a vast collection of straight and curved lines, which it converts to pixels. It takes an enormous amount of computer time to make sense out of this jumble of lines. It might take us a fraction of a second to recognize a table, but a computer sees only a collection of circles, ovals, spirals, straight lines, curly lines, corners, and so forth. After an enormous amount of computing time, a robot might finally recognize the object as a table. But if you rotate the image, the computer has to start all over again. In other words, robots can see, and in fact they can see much better than humans, but they don't understand what they are seeing. Upon entering a room, a robot would see only a jumble of lines and curves, not chairs, tables, and lamps.

Our brain unconsciously recognizes objects by performing trillions upon trillions of calculations when we walk into a room–an activity that we are blissfully unaware of. The reason that we are unaware of all our brain is doing is evolution. If we were alone in the forest with a charging saber-toothed tiger, we would be paralyzed if we were aware of all the computations necessary to recognize the danger and escape. For the sake of survival, all we need to know is how to run. When we lived in the jungle, it simply was not necessary for us to be aware of all of the ins and outs of our brain's recognizing the ground, the sky, the trees, the rocks, and so forth.

In other words, the way our brain works can be compared to a huge iceberg. We are aware of only the tip of the iceberg, the conscious mind. But lurking below the surface, hidden from view, is a much larger object, the unconscious mind, which consumes vast amounts of the brain's "computer power" to understand simple things surrounding it, such as figuring out where you are, whom you are talking to, and what lies around you. All this is done automatically without our permission or knowledge.

This is the reason that robots cannot navigate across a room, read handwriting, drive trucks and cars, pick up garbage, and so forth. The U.S. military has spent hundreds of millions of dollars trying to develop mechanical soldiers and intelligent trucks, without success.

Scientists began to realize that playing chess or multiplying huge numbers required only a tiny, narrow sliver of human intelligence. When the IBM computer Deep Blue beat world chess champion Garry Kasparov in a six-game match in 1997, it was a victory of raw computer power, but the experiment told us nothing about intelligence or consciousness, although the game made plenty of headlines. As Douglas Hofstadter, a computer scientist at Indiana University, said, "My God, I used to think chess required thought. Now, I realize it doesn't. It doesn't mean Kasparov isn't a deep thinker, just that you can bypass deep thinking in playing chess, the way you can fly without flapping your wings."

(Developments in computers will also have an enormous impact on the future of the job market. Futurists sometimes speculate that the only people who will have jobs decades into the future will be highly skilled computer scientists and technicians. But actually workers such as sanitation men, construction workers, firemen, police, and so forth, will also have jobs in the future because what they do involves pattern recognition. Every crime, piece of garbage, tool, and fire is different and hence cannot be managed by robots. Ironically, college-educated workers, such as low-level accountants, brokers, and tellers, may lose their jobs in the future since their work is semirepetitive and involves keeping track of numbers, a task that computers excel at.)

In addition to pattern recognition, the second problem with the development of robots is even more fundamental, and that is their lack of "common sense." Humans know, for example,

- Water is wet.
- Mothers are older than their daughters.
- Animals do not like pain.
- You don't come back after you die.
- Strings can pull, but not push.
- Sticks can push, but cannot pull.
- Time does not run backward.

But there is no line of calculus or mathematics that can express these truths. We know all of this because we have seen animals, water, and strings, and we have figured out the truth by ourselves. Children learn common sense by bumping into reality. The intuitive laws of biology and physics are learned the hard way, by interacting with the real world. But robots haven't experienced this. They know only what has been programmed into them beforehand.

(As a result, the jobs of the future will also include those that require common sense, that is, artistic creativity, originality, acting talent, humor, entertainment, analysis, and leadership. These are precisely the qualities that make us uniquely human and that computers have difficulty duplicating.)

In the past, mathematicians have tried to mount a crash program that could amass all the laws of common sense once and for all. The most ambitious attempt is CYC (short for encyclopedia), the brainchild of Douglas Lenat, the head of Cycorp. Like the Manhattan Project, the $2 billion crash program that built the atomic bomb, CYC was to be the "Manhattan Project" of artificial intelligence, the final push that would achieve true artificial intelligence.

Not surprisingly, Lenat's motto is, Intelligence is 10 million rules. (Lenat has a novel way in which to find new laws of common sense; he has his staff read the pages of scandalous tabloids and lurid gossip rags. Then he asks CYC if it can spot the errors in the tabloids. Actually, if Lenat succeeds in this, CYC may actually be more intelligent than most tabloid readers!)

One of the goals of CYC is to attain "breakeven," that is, the point at which a robot will be able to understand enough so that it can digest new information on its own simply by reading magazines and books found in any library. At that point, like a baby bird leaving the nest, CYC will be able to flap its wings and take off on its own.

But since the firm's founding in 1984, its credibility has suffered from a common problem in AI: making predictions that generate headlines but are wildly unrealistic. Lenat predicted that in ten years, by 1994, CYC would contain 30 to 50 percent of "consensus reality." Today CYC is not even close. As the scientists of Cycorp have found out, millions and millions of lines of code need to be programmed in order for a computer to approximate the common sense of a four-year-old child. So far the latest version of the CYC program contains only a paltry 47,000 concepts and 306,000 facts. Despite Cycorp's regularly optimistic press releases, one of Lenat's coworkers, R. V. Guha, who left the team in 1994, was quoted as saying, "CYC is generally viewed as a failed project.... We were killing ourselves trying to create a pale shadow of what had been promised."

In other words, attempts to program all the laws of common sense into a single computer have floundered, simply because there are so many laws of common sense. Humans learn these laws effortlessly because we tediously continue to bump into the environment throughout our lives, quietly assimilating the laws of physics and biology, but robots do not.

Microsoft founder Bill Gates admits, "It has been much harder than expected to enable computers and robots to sense their surrounding environment and to react quickly and accurately ... for example, the abilities to orient themselves with respect to the objects in a room, to respond to sounds and interpret speech, and to grasp objects of varying sizes, textures, and fragility. Even something as simple as telling the difference between an open door and a window can be devilishly tricky for a robot."

Proponents of the top-down approach to artificial intelligence, however, point out that progress in this direction, although at times glacial, is happening in labs around the world. For example, for the past few years the Defense Advanced Research Projects Agency (DARPA), which often funds state-of-the-art technology projects, has sponsored a $2 million prize for the creation of a driverless vehicle that can navigate by itself around a rugged terrain in the Mojave Desert. In 2004 not a single entry in the DARPA Grand Challenge could finish the race. In fact the top car managed to travel 7.4 miles before breaking down. But in 2005 the Stanford Racing Team's driverless car successfully navigated the grueling 132-mile course (although it took the car seven hours to do so). Four other cars also completed the race. (Some critics noted that the rules permitted the cars to use GPS navigation systems along a long deserted path; in effect, the cars could follow a predetermined road map without many obstructions, so the cars never had to recognize complex obstacles in their path. In real driving, cars have to navigate unpredictably around other cars, pedestrians, construction sites, traffic jams, and so forth.)

Bill Gates is cautiously optimistic that robotic machines may be the "next big thing." He likens the field of robotics now to the personal computer field he helped to start thirty years ago. Like the PC, it may be poised to take off. "No one can say with any certainty when–or if–this industry will achieve critical mass," he writes. "If it does, though, it may well change the world."

(Once robots with humanlike intelligence become commercially available, there will be a huge market for them. Although true robots do not exist today, preprogrammed robots do exist and have proliferated. The International Federation of Robotics estimates that there were about 2 million of these personal robots in 2004, and that another 7 million would be installed by 2008. The Japanese Robot Association predicts that by 2025 the personal robot industry, today worth $5 billion, will be worth $50 billion per year.)

The Bottom-Up Approach

Because of the limitations of the top-down approach to artificial intelligence, attempts have been made to use a "bottom-up" approach instead, that is, to mimic evolution and the way a baby learns. Insects, for example, do not navigate by scanning their environment and reducing the image to trillions upon trillions of pixels that they process with supercomputers. Instead insect brains are composed of "neural networks," learning machines that slowly learn how to navigate in a hostile world by bumping into it. At MIT, walking robots were notoriously difficult to create via the top-down approach. But simple buglike mechanical creatures that bump into the environment and learn from scratch can successfully scurry around the floor at MIT within a matter of minutes.

Rodney Brooks, director of MIT's famed Artificial Intelligence Laboratory, famous for its huge, lumbering "top-down" walking robots, became a heretic when he explored the idea of tiny "insectoid" robots that learned to walk the old-fashioned way, by stumbling and bumping into things. Instead of using elaborate computer programs to mathematically compute the precise position of their feet as they walked, his insectoids used trial and error to coordinate their leg motions using little computer power. Today many of the descendants of Brooks's insectoid robots are on Mars gathering data for NASA, scurrying across the bleak Martian landscape with a mind of their own. Brooks believes that his insectoids are ideally suited to explore the solar system.

One of Brooks's projects has been COG, an attempt to create a mechanical robot with the intelligence of a six-month-old child. On the outside COG looks like a jumble of wires, circuits, and gears, except that it has a head, eyes, and arms. No laws of intelligence have been programmed into it. Instead it is designed to focus its eyes on a human trainer, who tries to teach it simple skills. (One researcher who became pregnant made a bet as to which would learn faster, COG or her child by the age of two. The child far surpassed COG.)

For all the successes in mimicking the behavior of insects, robots using neural networks have performed miserably when their programmers have tried to duplicate in them the behavior of higher organisms like mammals. The most advanced robot using neural networks can walk across the room or swim in water, but it cannot jump and hunt like a dog in the forest, or scurry around the room like a rat. Many large neural network robots may consist of tens to perhaps hundreds of "neurons"; the human brain, however, has over 100 billion neurons. *C. elegans,* a very simple worm whose nervous system has been completely mapped by biologists, has just over 300 neurons in its nervous system, making its nervous system perhaps one of the simplest found in nature. But there are over 7,000 synapses between these neurons. As simple as *C. elegans* is, its nervous system is so complex that no one has yet been able to construct a computer model of this brain. (In 1988 one computer expert predicted that by now we should have robots with about 100 million artificial neurons. Actually, a neural network with 100 neurons is considered exceptional.)

The supreme irony is that machines can effortlessly perform tasks that humans consider "hard," such as multiplying large numbers or playing chess, but machines stumble badly when asked to perform tasks that are supremely "easy" for human beings, such as walking across a room, recognizing faces, or gossiping with a friend. The reason is that our most advanced computers are basically just adding machines. Our brain, however, is exquisitely designed by evolution to solve the mundane problems of survival, which require a whole complex architecture of thought, such as common sense and pattern recognition. Survival in the forest did not depend on calculus or chess, but on evading predators, finding mates, and adjusting to changing environments.

MIT's Marvin Minsky, one of the original founders of AI, summarizes the problems of AI in this way: "The history of AI is sort of funny because the first real accomplishments were beautiful things, like a machine that could do proofs in logic or do well in a calculus course. But then we started to try to make machines that could answer questions about the simple kinds of stories that are in a first-grade reader book. There's no machine today that can do that."

Some believe that eventually there will be a grand synthesis between the two approaches, the top-down and bottom-up, which may provide the key to artificial intelligence and humanlike robots. After all, when a child learns, although he first relies mainly on the bottom-up approach, bumping into his surroundings, eventually he receives instruction from parents, books, and schoolteachers, and learns from the top-down approach. As an adult, we constantly blend these two approaches. A cook, for example, reads from a recipe but also constantly samples the dish as it is cooking.

Hans Moravec says, "Fully intelligent machines will result when the mechanical golden spike is driven uniting the two efforts," probably within the next forty years.

Emotional Robots?

One consistent theme in literature and art is the mechanical being that yearns to become human, to share in human emotions. Not content to be made of wires and cold steel, it wishes to laugh, cry, and feel all the emotional pleasures of a human being.

Pinocchio, for example, was the puppet that wanted to become a real boy. The Tin Man in the *The Wizard of Oz* wanted to have a heart. And Data, on *Star Trek,* is a robot that can outperform all humans in strength and intelligence, yet still yearns to become human.

Some people have even suggested that our emotions represent the highest quality of what it means to be human. No machine will ever be able to thrill at a blazing sunset or laugh at a humorous joke, they claim. Some say that it is impossible for machines ever to have emotions, since emotions represent the pinnacle of human development.

But the scientists working on AI and trying to break down emotions paint a different picture. To them emotions, far from being the essence of humanity, are actually a by-product of evolution. Simply put, emotions are good for us. They helped us to survive in the forest, and even today they help us to navigate the dangers of life.

For example, "liking" something is very important evolutionarily, because most things are harmful to us. Of the millions of objects that we bump into every day, only a handful are beneficial to us. Hence to "like" something is to make a distinction between one out of the tiny fraction of things that can help us over against the millions of things that might hurt us.

Similarly, jealousy is an important emotion, because our reproductive success is vital in ensuring the survival of our genes to the next generation. (In fact, that is why there are so many emotionally charged feelings related to sex and love.)

Shame and remorse are important because they help us to learn the socialization skills necessary to function in a cooperative society. If we never say we're sorry, eventually we will be expelled from the tribe, diminishing our chances of surviving and passing on our genes.

Loneliness, too, is an essential emotion. At first loneliness seems to be unnecessary and redundant. After all, we can function alone. But longing to be with companions is also important for our survival, since we depend on the resources of the tribe to survive.

In other words, when robots become more advanced, they, too, might be equipped with emotions. Perhaps robots will be programmed to bond with their owners or caretakers, to ensure that they don't wind up in the garbage dump. Having such emotions would help to ease their transition into society, so that they could be helpful companions, rather than rivals of their owners.

Computer expert Hans Moravec believes that robots will be programmed with emotions such as "fear" to protect themselves. For example, if a robot's batteries are running down, the robot "would express agitation, or even panic, with signals that humans can recognize. It would go to the neighbors and ask them to use their plug, saying, 'Please! Please! I need this! It's so important, it's such a small cost! We'll reimburse you!' "

Emotions are vital in decision making, as well. People who have suffered a certain kind of brain injury lack the ability to experience emotions. Their reasoning ability is intact, but they cannot express any feelings. Neurologist Dr. Antonio Damasio of the University of Iowa College of Medicine, who has studied people with these types of brain injuries, concludes that they seem "to know, but not to feel."

Dr. Damasio finds that such individuals are often paralyzed in making the smallest decisions. Without emotions to guide them, they endlessly debate over this option or that option, leading to crippling indecision. One patient of Dr. Damasio spent half an hour trying to decide the date of his next appointment.

Scientists believe that emotions are processed in the "limbic system" of the brain, which lies deep in the center of our brain. When people suffer from a loss of communication between the neocortex (which governs rational thinking) and the limbic system, their reasoning powers are intact but they have no emotions to guide them in making decisions. Sometimes we have a "hunch" or a "gut reaction" that propels our decision making. People with injuries that effect the communication between the rational and emotional parts of the brain do not have this ability.

For example, when we go shopping we unconsciously make thousands of value judgments about almost everything we see, such as "This is too expensive, too cheap, too colorful, too silly, or just right." For people with this type of brain injury, shopping can be a nightmare because everything seems to have the same value.

As robots become more intelligent and are able to make choices of their own, they could likewise become paralyzed with indecision. (This is reminiscent of the parable of the donkey sitting between two bales of hay that eventually dies of starvation because it cannot decide which to eat.) To aid them, robots of the future may need to have emotions hardwired into their brains. Commenting on the lack of emotions in robots, Dr. Rosalind Picard of the MIT Media Lab says, "They can't feel what's most important. That's one of their biggest failings. Computers just don't get it."

As Russian novelist Fyodor Dostoevsky wrote, "If everything on Earth were rational, nothing would happen."

In other words, robots of the future may need emotions to set goals and to give meaning and structure to their "lives," or else they will find themselves paralyzed with infinite possibilities.

Are They Conscious?

There is no universal consensus as to whether machines can be conscious, or even a consensus as to what consciousness means. No one has come up with a suitable definition of consciousness.

Marvin Minsky describes consciousness as more of a "society of minds," that is, the thinking process in our brain is not localized but spread out, with different centers competing with one another at any given time. Consciousness may then be viewed as a sequence of thoughts and images issuing from these different, smaller "minds," each one grabbing and competing for our attention.

If this is true, perhaps "consciousness" has been overblown, perhaps there have been too many papers devoted to a subject that has been overmystified by philosophers and psychologists. Maybe defining consciousness is not so hard. As Sydney Brenner of the Salk Institute in La Jolla says, "I predict that by 2020–the year of good vision–consciousness will have disappeared as a scientific problem. . . . Our successors will be amazed by the amount of scientific rubbish discussed today–that is, if they have the patience to trawl through the electronic archives of obsolete journals."

AI research has been suffering from "physics envy," according to Marvin Minsky. In physics the holy grail has been to find a simple equation that will unify the physical forces of the universe into a single theory, creating a "theory of everything." AI researchers, overly influenced by this idea, have tried to find a single paradigm that would explain consciousness. But such a simple paradigm may not exist, according to Minsky.

(Those in the "constructionist" school, like myself, believe that instead of endlessly debating whether thinking machines can be created or not, one should instead try to build one. Regarding consciousness, there is probably a continuum of consciousness, from a lowly thermostat that monitors the temperature in a room to the self-aware organisms that we are today. Animals may be conscious, but they do not possess the level of consciousness of a human being. One should try, therefore, to categorize all the various types and levels of consciousness rather than debate philosophical questions about the meaning of consciousness. Robots may eventually attain a "silicon consciousness." Robots, in fact, may one day embody an architecture for thinking and processing information that is different from ours. In the future, advanced robots might blur the difference between syntax and semantics, so that their responses will be indistinguishable from the responses of a human. If so, the question of whether they really "understand" the question will be largely irrelevant. A robot that has perfect mastery of syntax, for all practical purposes, understands what is being said. In other words, a perfect mastery of syntax *is* understanding.)

Could Robots Be Dangerous?

Because of Moore's law, which states that computer power doubles every eighteen months, it is conceivable that within a few decades robots will be created that have the intelligence, say, of a dog or a cat. But by 2020 Moore's law may well collapse and the age of silicon could come to an end. For the past fifty years or so the astounding growth in computer power has been fueled by the ability to create tiny silicon transistors, tens of millions of which can easily fit on your fingernail. Beams of ultraviolet radiation are used to etch microscopic transistors onto wafers made of silicon. But this process cannot last forever. Eventually, these transistors could become so small that they reach the size of molecules, and the process will break down. Silicon Valley could become a Rust Belt after 2020, when the age of silicon finally comes to an end.

The Pentium chip in your laptop computer has a layer about twenty atoms across. By 2020 that Pentium chip might consist of a layer only five atoms across. At that point the Heisenberg uncertainty principle kicks in, and you no longer know where the electron is. Electricity will then leak out of the chip and the computer will short-circuit. At that point, the computer revolution and Moore's law will hit a dead end because of the laws of the quantum theory. (Some people have claimed that the digital era is the "victory of bits over atoms." But eventually, when we hit the limit of Moore's law, atoms may have their revenge.)

Physicists are now working on the post-silicon technology that will dominate the computer world after 2020, but so far with mixed results. As we have seen, a variety of technologies are being studied that may eventually replace silicon technology, including quantum computers, DNA computers, optical computers, atomic computers, and so forth. But each of them faces huge hurdles before it can take on the mantle of silicon chips. Manipulating individual atoms and molecules is a technology that is still in its infancy, so making billions of transistors that are atomic in size is still beyond our ability.

But assume, for the moment, that physicists are capable of bridging the gap between silicon chips and, say, quantum computers. And assume that some form of Moore's law continues into the post-silicon era. Then artificial intelligence might become a true possibility. At that point robots might master human logic and emotions and pass the Turing test every time. Steven Spielberg explored this question in his movie *Artificial Intelligence: AI,* where the first robot boy was created that could exhibit emotions, and was hence suitable for adoption into a human family.

This raises the question: could such robots be dangerous? The answer is likely yes. They could become dangerous once they have the intelligence of a monkey, which is self-aware and can create its own agenda. It may take many decades to reach such a point, so scientists will have plenty of time to observe robots before they pose a threat. For example, a special chip could be placed in their processors that could prevent them from going on the rampage. Or they could have a self-destruct or deactivation mechanism that would turn them off in case of an emergency.

Arthur C. Clarke wrote, "It is possible that we may become pets of the computers, leading pampered existences like lapdogs, but I hope that we will always retain the ability to pull the plug if we feel like it."

A more mundane threat is that our infrastructure depends on computers. Our water and electricity grid, not to mention transportation and communications networks, will be increasingly computerized in the future. Our cities have become so complex that only complex and intricate computer networks can regulate and monitor our vast infrastructure. In the future it will become increasingly important to add artificial intelligence to this computer network. A failure or breakdown in this all-pervasive computer infrastructure could paralyze a city, country, or even a civilization.

Will computers eventually surpass us in intelligence? Certainly, there is nothing in the laws of physics to prevent that. If robots are neural networks capable of learning, and they develop to the point where they can learn faster and more efficiently than we can, then it's logical that they might eventually surpass us in reasoning. Moravec says, "[The postbiological world] is a world in which the human race has been swept away by the tide of cultural change, usurped by its own artificial progeny . . . When that happens, our DNA will find itself out of a job, having lost the evolutionary race to a new kind of competition."

Some inventors, such as Ray Kurzweil, have even predicted that this time will come soon, earlier rather than later, even within the next few decades. Perhaps we are creating our evolutionary successors. Some computer scientists envision a point they call "singularity," when robots will be able to process information exponentially fast, creating new robots in the process, until their collective ability to absorb information advances almost without limit.

So in the long term some have advocated a merging of carbon and silicon technology, rather than waiting for our extinction. We humans are mainly based on carbon, but robots are based on silicon (at least for the moment). Perhaps the solution is to merge with our creations. (If we ever encounter extraterrestrials, we should not be surprised to find that they are part organic, part mechanical to withstand the rigors of space travel and to flourish in hostile environments.)

In the far future, robots or humanlike cyborgs may even grant us the gift of immortality. Marvin Minsky adds, "What if the sun dies out, or we destroy the planet? Why not make better physicists, engineers, or mathematicians? We may need to be the architects of our own future. If we don't our culture could disappear."

Moravec envisions a time in the distant future when our neural architecture will be transferred, neuron for neuron, directly into a machine, giving us, in a sense, immortality. It's a wild thought, but not beyond the realm of possibility. So, according to some scientists viewing the far future, immortality (in the form of DNA-enhanced or silicon bodies) may be the ultimate future of humanity.

The idea of creating thinking machines that are at least as smart as animals and perhaps as smart or smarter than us could become a reality if we can overcome the collapse of Moore's law and the commonsense problem, perhaps even late in this century. Although the fundamental laws of AI are still being discovered, progress in this area is happening extremely fast and is promising. Given that, I would classify robots and other thinking machines as a Class I impossibility.

8: EXTRATERRESTRIALS AND UFOS

Either we are alone in the universe, or we are not.
Either thought is frightening.
–ARTHUR C. CLARKE

A gargantuan spaceship, stretching miles across, looms directly over Los Angeles, filling up the entire sky and ominously darkening the entire city. All over the world, saucer-shaped fortresses position themselves over the major cities of the world. Hundreds of jubilant spectators, wishing to welcome the beings from another planet to L.A., gather on top of a skyscraper to reach out to their celestial guests.

After days of hovering silently over L.A., the spaceship's belly slowly opens up. A searing blast of laser light shoots out, incinerating the skyscraper, unleashing a tidal wave of destruction that rolls across the entire city, reducing it to burned rubble within seconds.

In the movie *Independence Day* aliens represent our deepest fears. In the movie *E.T.* we project onto aliens our own dreams and fantasies. Throughout history people have been fascinated by the thought of alien creatures that inhabit other worlds. As far back as 1611, in his treatise *Somnium,* the astronomer Johannes Kepler, using the best scientific knowledge of the time, speculated about a trip to the moon during which one might encounter strange aliens, plants, and animals. But science and religion often collide on the subject of life in space, sometimes with tragic results.

A few years earlier, in 1600, former Dominican monk and philosopher Giordano Bruno was burned alive in the streets of Rome. To humiliate him, the Church hung him upside down and stripped him naked before finally burning him at the stake. What made the teachings of Bruno so dangerous? He had asked a simple question: is there life in outer space? Like Copernicus, he believed that the Earth revolved around the sun, but unlike Copernicus, he believed that there could be countless numbers of creatures like us living in outer space. (Rather than entertain the possibility of billions of saints, popes, churches, and Jesus Christs in outer space, it was more convenient for the Church simply to burn him.)

For four hundred years the memory of Bruno has haunted the historians of science. But today Bruno has his revenge every few weeks. About twice a month a new extrasolar planet is discovered orbiting a star in space. Over 250 planets have now been documented orbiting other stars in space. Bruno's prediction of extrasolar planets has been vindicated. But one question still lingers. Although the Milky Way galaxy may be teaming with extrasolar planets, how many of them can support life? And if intelligent life does exist in space, what can science say about it?

Hypothetical encounters with extraterrestrials, of course, have fascinated society and thrilled readers and movie audiences for generations. The most famous incident occurred on October 30, 1938, when Orson Welles decided to play a Halloween trick on the American public. He took the basic plot of H. G. Wells's *War of the Worlds* and made a series of short news announcements on CBS national radio, interrupting dance music to reenact, hour by hour, the invasion of Earth by Martians and the subsequent collapse of civilization. Millions of Americans were panic-stricken over the "news" that machines from Mars had landed in Grover's Mill, New Jersey, and were unleashing death rays to destroy entire cities and conquer the world. (Newspapers later recorded that spontaneous evacuations took place as people fled the area, with eyewitnesses claiming they could smell poison gas and see flashes of light in the distance.)

Fascination with Mars peaked again in the 1950s, when astronomers noticed a strange marking on Mars that looked like a gigantic *M* that was hundreds of miles across. Commentators noted that perhaps the *M* stood for "Mars," and Martians were peacefully signaling their presence to earthlings, like cheerleaders spelling out their team's name in a football stadium. (Others noted darkly that the *M* marking was actually a *W*, and *W* stands for "war." In other words, the Martians were actually declaring war on the Earth!) The mini-panic eventually subsided when this mysterious *M* disappeared just as abruptly as it had appeared. In all likelihood this marking was caused by a dust storm that covered the entire planet, except for the tops of four large volcanoes. The tops of these volcanoes roughly took on the shape of an *M* or a *W*.

THE SCIENTIFIC SEARCH FOR LIFE

Serious scientists studying the possibility of extraterrestrial life state that it is impossible to say anything definitive about such life, assuming that it exists. Nonetheless, we can make some general arguments on the nature of alien life based on what we know of physics, chemistry, and biology.

First, scientists believe that liquid water will be the key factor in creating life in the universe. "Follow the water" is the mantra recited by astronomers as they search for evidence of life in space. Liquid water, unlike most liquids, is a "universal solvent" that can dissolve an astonishing variety of chemicals. It is an ideal mixing bowl to create increasingly complex molecules. Water is also a simple molecule that is found throughout the universe, while other solvents are quite rare.

Second, we know that carbon is a likely component in creating life because it has four bonds and hence the ability to bind to four other atoms and create molecules of incredible complexity. In particular, it is easy to form long carbon chains, which become the basis for hydrocarbon and organic chemistry. Other elements with four bonds do not have such a rich chemistry.

The most vivid illustration of the importance of carbon was the famous experiment conducted by Stanley Miller and Harold Urey in 1953, which showed that the spontaneous formation of life may be a natural by-product of carbon chemistry. They took a solution of ammonia, methane, and other toxic chemicals that they believed were found in the early Earth, put it in a flask, exposed it to a small electrical current, and then simply waited. Within one week they could see evidence of amino acids forming spontaneously in the flask. The electrical current was sufficient to break apart the carbon bonds within ammonia and methane and then rearrange the atoms into amino acids, the precursors of proteins. In some sense, life can form spontaneously. Since then, amino acids have been found inside meteorites and also in gas clouds in deep space.

Third, the fundamental basis of life is the self-replicating molecule called DNA. In chemistry, self-replicating molecules are extremely rare. It took hundreds of millions of years to form the first DNA molecule on Earth, probably deep in the oceans. Presumably, if one could perform the Miller-Urey experiment for a million years in the oceans, DNA-like molecules would spontaneously form. One likely site where the first DNA molecule on Earth might have occurred early in the Earth's history is near volcano vents on the ocean bottom, since the activity of the vents would create a convenient supply of energy for the early DNA molecule and cells, before the arrival of photosynthesis and plants. It is not known if other carbon-based molecules besides DNA can also be self-replicating, but it is likely that other self-replicating molecules in the universe will resemble DNA in some way.

So life probably requires liquid water, hydrocarbon chemicals, and some form of self-replicating molecule like DNA. Using these broad criteria one can derive a rough estimate for the frequency of intelligent life in the universe. In 1961 Cornell University astronomer Frank Drake was one of the first to make a rough estimate. If you start with 100 billion stars in the Milky Way galaxy, you can estimate what fraction of them have stars like our sun. Of these, you can estimate what fraction have solar systems revolving around them.

More specifically, Drake's equation calculates the number of civilizations in the galaxy by multiplying several numbers together, including

- the rate at which stars are born in the galaxy,
- the fraction of these stars that have planets,
- the number of planets for each star that have the conditions for life,
- the fraction of planets that actually develop life,
- the fraction that develop intelligent life,
- the fraction that are willing and able to communicate, and
- the expected lifetime of a civilization.

By taking reasonable estimates and by multiplying these successive probabilities, one realizes that there could be between 100 and 10,000 planets in the Milky Way galaxy alone that are able to harbor intelligent life. If these intelligent life-forms are uniformly scattered across the Milky Way galaxy, then we should expect to find such a planet just a few hundred light-years from Earth. In 1974 Carl Sagan estimated that there might be up to a million such civilizations within our Milky Way galaxy alone.

This theorizing, in turn, has provided added justification for those looking to find evidence for extraterrestrial civilizations. Given the favorable estimate of planets capable of harboring intelligent life-forms, scientists have begun seriously to look for the radio signals such planets may have emitted, much like the TV and radio signals that our own planet has been emitting for the past fifty years.

Listening to ET

The Search for Extraterrestrial Intelligence (SETI) project dates back to an influential paper written in 1959 by physicists Giuseppe Cocconi and Philip Morrison, who suggested that listening to microwave radiation of a frequency between 1 and 10 gigahertz would be the most suitable way to eavesdrop on extraterrestrial communications. (Below 1 gigahertz, signals would be washed out by radiation emitted by fast-moving electrons; beyond 10 gigahertz, noise from oxygen and water molecules in our own atmosphere would interfere with any signals.) They selected 1,420 gigahertz as the most promising frequency in which to listen to signals from outer space, since that was the emission frequency for ordinary hydrogen gas, the most plentiful element in the universe. (Frequencies around that range are nicknamed the "watering hole," given their convenience for extraterrestrial communication.)

Searches for evidence of intelligent signals near the watering hole, however, have been disappointing. In 1960 Frank Drake initiated Project Ozma (named after the Queen of Oz) to search for signals using the 25-meter radio telescope in Green Bank, West Virginia. No signals were ever found, either in Project Ozma or in other projects that, in fits and starts, tried to scan the night sky over the years.

In 1971 an ambitious proposal was made by NASA to fund SETI research. Dubbed Project Cyclops, the effort involved fifteen hundred radio telescopes at a cost of $10 billion. Not surprisingly, the research never went anywhere. Funding did become available for a much more modest proposal, to send a carefully coded message to alien life in outer space. In 1974 a coded message of 1,679 bits was transmitted via the giant Arecibo radio telescope in Puerto Rico toward the Globular Cluster M13, about 25,100 light-years away. In this short message, scientists created a 23 × 73 dimensional grid pattern that plotted the location of our solar system, containing an illustration of human beings and some chemical formulae. (Because of the large distances involved, the earliest date for a reply from outer space would be 52,174 years from now.)

Congress has not been impressed with the significance of these projects, even after a mysterious radio signal, called the "Wow" signal, was received in 1977. It consisted of a series of letters and numbers that seemed to be nonrandom and seemed to be signaling the existence of intelligence. (Some who have seen the Wow signal have not been convinced.)

In 1995, frustrated by the lack of funding from the federal government, astronomers turned to private sources to start the nonprofit SETI

Institute in Mountain View, California, to centralize SETI research and initiate Project Phoenix to study one thousand nearby sunlike stars in the 1,200- to 3,000-megahertz range. Dr. Jill Tarter (the model for the scientist played by Jodie Foster in the movie *Contact*) was named director. (The equipment used in the project was so sensitive that it could pick up the emissions from an airport radar system 200 light-years away.)

Since 1995 the SETI Institute has scanned more than one thousand stars at a cost of $5 million per year. But there have been no tangible results. Nevertheless, Seth Shostak, senior astronomer at SETI, optimistically believes that the 350-antenna Allen Telescope Array now being built 250 miles northeast of San Francisco "will trip across a signal by the year 2025."

A more novel approach is the SETI@home project, initiated by astronomers at the University of California at Berkeley in 1999. They hit upon the idea of enlisting millions of PC owners whose computers sit idle most of the time. Those who participate download a software package that will help to decode some of the radio signals received by a radio telescope while the participant's screen saver is activated, so there is no inconvenience to the PC user. So far the project has signed up 5 million users in more than two hundred countries, consuming over a billion dollars of electricity, all at little cost. It is the most ambitious collective computer project ever undertaken in history and could serve as a model for other projects that need vast computer resources to carry out computations. So far no signal from an intelligent source has been found by SETI@home.

After decades of hard work, the glaring lack of any progress in SETI research has forced its proponents to ask hard questions. One obvious defect might be the exclusive use of radio signals at certain frequency bands. Some have suggested that alien life might use laser signals instead of radio signals. Lasers have several advantages over radio, because a laser's short wavelength means that you can pack more signals into one wave than you can with radio. But because laser light is highly directional and also contains just one frequency, it is exceptionally hard to tune into precisely the right laser frequency.

Another obvious defect might be SETI researchers' reliance on certain radio frequency bands. If there is alien life, it may use compression techniques or might disperse messages via smaller packages, strategies that are used on the modern Internet today. Listening in on compressed messages that have been spread over many frequencies, we might hear only random noise.

But given all the formidable problems facing SETI, it is reasonable to assume that sometime in this century we should be able to detect some signal from an extraterrestrial civilization, assuming that such civilizations exist. And should that happen, it would represent a milestone in the history of the human race.

WHERE ARE THEY?

The fact that the SETI project so far has found no indication of signals from intelligent life in the universe has forced scientists to take a cold, hard look at the assumptions behind Frank Drake's equations for intelligent life on other planets. Recently astronomical discoveries have led us to believe that the chance of finding intelligent life are much different than originally computed by Drake in the 1960s. The chance that intelligent life exists in the universe is both more optimistic and more pessimistic than originally believed.

First, new discoveries have led us to believe that life can flourish in ways not considered by Drake's equations. Before, scientists believed that liquid water can exist only in the "Goldilocks zone" surrounding the sun. (The distance from Earth to the sun is "just right." Not too close to the sun, because the oceans would boil, and not too far away, because the oceans would freeze, but "just right" to make life possible.)

So it came as a shock when astronomers found evidence that liquid water may exist beneath the ice cover on Europa, a frozen moon of Jupiter. Europa is well outside the Goldilocks zone, so it would appear not to fit the conditions of Drake's equation. Yet tidal forces might be sufficient to melt the ice cover of Europa and produce a permanent liquid ocean. As Europa spins around Jupiter, the planet's huge gravitational field squeezes the moon like a rubber ball, creating friction deep within its core, which in turn could cause the ice cover to melt. Since there are over one hundred moons in our solar system alone, this means that there could be an abundance of life-supporting moons in our solar system outside the Goldilocks zone. (And the 250 or so giant extrasolar planets so far discovered in space might also have frozen moons that can support life.)

Furthermore, scientists believe the universe could be peppered with wandering planets that no longer circle around any star. Because of tidal forces, any moon orbiting a wandering planet might have liquid oceans under its ice cover and hence life, but such moons would be impossible to see by our instruments, which depend on detecting light from the mother star.

Given that the number of moons probably greatly outnumbers the number of planets in a solar system, and given that there could be millions of wandering planets in the galaxy, the number of astronomical bodies with life-forms in the universe might be much larger than previously believed.

On the other hand, other astronomers have concluded, for a variety of reasons, that the chances for life on planets within the Goldilocks zone are probably much lower than originally estimated by Drake.

First, computer programs show that the presence of a Jupiter-sized planet in a solar system is necessary to fling passing comets and meteors into space, thereby continually cleaning out a solar system and making life possible. If Jupiter did not exist in our solar system, Earth would be pelted with meteors and comets, making life impossible. Dr. George Wetherill, an astronomer at the Carnegie Institution in Washington, D.C., estimates that without the presence of Jupiter or Saturn in our solar system, the Earth would have suffered a thousand times more asteroid collisions, with a huge life-threatening impact (like the one that destroyed the dinosaurs 65 million years ago) occurring every ten thousand years. "It's hard to imagine how life could survive that extreme onslaught," he says.

Second, our planet is blessed with a large moon, which helps to stabilize the Earth's spin. Extending Newton's laws of gravity over millions of years, scientists can show that without a large moon, our Earth's axis probably would have become unstable and the Earth might have tumbled, making life impossible. French astronomer Dr. Jacques Lasker estimates that without our moon the Earth's axis could oscillate between 0 and 54 degrees, which would precipitate extreme weather conditions incompatible with life. So the presence of a large moon also has to be factored into conditions used for Drake's equations. (The fact that Mars has two tiny moons, too small to stabilize its spin, means that Mars may have tumbled in the distant past, and may tumble again in the future.)

Third, recent geological evidence points to the fact that many times in the past, life on Earth was almost extinguished. About 2 billion years ago the Earth was probably completely covered in ice; it was a "snowball Earth" that could barely support life. At other times, volcanic eruptions and meteor impacts might have come close to destroying all life on Earth. So the creation and evolution of life is more fragile than we originally thought.

Fourth, intelligent life was also nearly extinguished in the past. About a hundred thousand years ago there were probably only a few hundred to a few thousand humans, based on the latest DNA evidence. Unlike most animals within a given species, which are separated by large genetic differences, humans are all nearly alike genetically. Compared to the animal kingdom, we are almost like clones of each other. This phenomenon can only be explained if there were "bottlenecks" in our history in which most of the human race was nearly wiped out. For example, a large volcanic eruption might have caused the weather to suddenly get cold, nearly killing off the entire human race.

There are still other fortuitous accidents that were necessary to spawn life on Earth, including

- *A strong magnetic field.* This is necessary in order to deflect cosmic rays and radiation that could destroy life on Earth.
- *A moderate speed of planetary rotation.* If the Earth rotated too slowly, the side facing the sun would be blisteringly hot, while the

other side would be freezing cold for long periods of time; if the Earth rotated too quickly, there would be extremely violent weather conditions, such as monster winds and storms.

- *A location that is the right distance from the center of the galaxy.* If the Earth were too close to the center of the Milky Way galaxy, it would be hit with dangerous radiation; if it were too far from the center, our planet would not have enough higher elements to create DNA molecules and proteins.

For all these reasons astronomers now believe that life might exist outside the Goldilocks zone on moons or wandering planets, but that the chances of the existence of a planet like Earth capable of supporting life within the Goldilocks zone are much lower than previously believed. Overall most estimates of Drake's equations show that the chances of finding civilization in the galaxy are probably smaller than he originally estimated.

As Professors Peter Ward and Donald Brownlee have written, "We believe that life in the form of microbes and their equivalents is very common in the universe, perhaps more common than even Drake and [Carl] Sagan envisioned. However, complex life–animals and higher plants–is likely to be far more rare than is commonly assumed." In fact, Ward and Brownlee leave open the possibility that the Earth may be unique in the galaxy for harboring animal life. (Although this theory may dampen the search for intelligent life in our galaxy, it still leaves open the possibility of life existing in other distant galaxies.)

The Search for Earth-like Planets

Drake's equation, of course, is purely hypothetical. That is why the search for life in outer space has gotten a boost from the discovery of extrasolar planets. What has hindered research into extrasolar planets is that they are invisible to any telescope since they give off no light of their own. They are in general a million to a billion times dimmer than the mother star.

To find them astronomers are forced to analyze tiny wobblings in the mother star, assuming that a large Jupiter-sized planet is capable of altering the orbit of the star. (Imagine a dog chasing its tail. In the same way, the mother star and its Jupiter-size planet "chase" each other by revolving around each other. A telescope cannot see the Jupiter-sized planet, which is dark, but the mother star is clearly visible and appears to wobble back and forth.)

The first true extrasolar planet was found in 1994 by Dr. Alexandr Wolszczan of Pennsylvania State University, who observed planets revolving around a dead star, a rotating pulsar. Because the mother star had probably exploded as a supernova, it seemed likely that these planets were dead, scorched planets. The following year two Swiss astronomers, Michel Mayor and Didier Queloz of Geneva, announced that they had found a more promising planet with a mass similar to Jupiter orbiting the star 51 Pegasi. Soon after that the floodgates were opened.

In the last ten years there has been a spectacular acceleration in the number of extrasolar planets being found. Geologist Bruce Jakosky of the University of Colorado at Boulder says, "This is a special time in the history of humanity. We're the first generation that has a realistic chance of discovering life on another planet."

None of the solar systems found so far resemble our own. In fact, they are all quite dissimilar to our solar system. Once, astronomers thought that our solar system was typical of others throughout the universe, with circular orbits and three rings of planets surrounding the mother star: a rocky belt of planets closest to the star, next a belt of gas giants, and finally a comet belt of frozen icebergs.

Much to their surprise, astronomers found that none of the planets in other solar systems followed that simple rule. In particular, Jupiter-sized planets were expected to be found far from the mother star, but instead many of them orbited either extremely close to the mother star (even closer than the orbit of Mercury) or in extremely elliptical orbits. Either way the existence of a small, Earth-like planet orbiting in the Goldilocks zone would be impossible in either condition. If the Jupiter-sized planet orbited too close to the mother star, it meant that the

Jupiter-sized planet had migrated from a great distance and gradually spiraled into the center of the solar system (probably due to friction caused by dust). In that case, the Jupiter-size planet would eventually cross the orbit of the smaller, Earth-like planet, flinging it into outer space. And if the Jupiter-sized planet followed a highly elliptical orbit, it would mean that the Jupiter-sized planet would pass regularly through the Goldilocks zone, again causing any Earth-like planet to be flung into space.

These findings were disappointing to planet hunters and astronomers hoping to discover other Earth-like planets, but in hindsight these findings were to be expected. Our instruments are so crude that they can detect only the largest, fastest-moving Jupiter-sized planet that can have a measurable effect on the mother star. Hence it is not surprising that today's telescopes can detect only monster planets that are moving rapidly in space. If an exact twin of our own solar system exists in outer space, our instruments are probably too crude to find it.

All this may change, with the launching of *Corot*, *Kepler*, and the *Terrestrial Planet Finder*, three satellites that are designed to locate several hundred Earth-like planets in space. The *Corot* and *Kepler* satellites, for example, will examine the faint shadow that would be cast by an Earth-like planet as it crosses the face of the mother star, slightly reducing its sunlight. Although the Earth-like planet would not be visible, the reduction in sunlight from the mother star could be detected by satellite.

The French *Corot* satellite (which in French stands for Convection, Stellar Rotation, and Planetary Transits) was successfully launched on December 2006 and represents a milestone, the first space-based probe to search for extrasolar planets. Scientists hope to find between ten and forty Earth-like planets. If they do, the planets will probably be rocky, not gas giants, and will be just a few times bigger than the Earth. *Corot* will also probably add to the many Jupiter-sized planets already found in space. "*Corot* will be able to find extrasolar planets of all sizes and natures, contrary to what we can do from the ground at the moment," says astronomer Claude Catala. Altogether scientists hope the satellite will scan up to 120,000 stars.

Any day, the *Corot* may find evidence of the first Earth-like planet in space, which will be a turning point in the history of astronomy. In the future people may have an existential shock gazing at the night sky and realizing that there are planets out there that could harbor intelligent life. When we look into the heavens in the future, we might find ourselves wondering if anyone is looking back.

The *Kepler* satellite is tentatively scheduled for launch in late 2008 by NASA. It is so sensitive that it may be able to detect up to hundreds of Earth-like planets in outer space. It will measure the brightness of 100,000 stars to detect the motion of any planet as it crosses the face of the star. During the four years it will be in operation, *Kepler* will analyze and monitor thousands of distant stars up to 1,950 light-years from Earth. In its first year in orbit, scientists expect the satellite to find roughly

- 50 planets about the same size as Earth,
- 185 planets about 30 percent larger than the Earth, and
- 640 planets about 2.2 times the size of the Earth.

The *Terrestrial Planet Finder* may have an even better chance of finding Earth-like planets. After several delays, it is tentatively scheduled for launch in 2014; it will analyze as many as one hundred stars up to 45 light-years away with great accuracy. It will be equipped with two separate devices to search for distant planets. The first is a coronagraph, a special telescope that blocks out the sunlight from the mother star, reducing its light by a factor of a billion. The telescope will be three to four times bigger than the Hubble Space Telescope and ten times more precise. The second device on the Finder is an interferometer, which uses the interference of light waves to cancel the light from the mother star by a factor of a million.

Meanwhile the European Space Agency is planning to launch its own planet finder, the *Darwin*, to be sent into orbit in 2015 or later. It is planned to consist of three space telescopes, each about 3 meters in diameter, flying in formation, acting as one large interferometer. Its mission, too, will be to identify Earth-like planets in space.

Identifying hundreds of Earth-like planets in outer space will help to refocus the SETI effort. Instead of randomly scanning nearby stars, astronomers will be able to pinpoint their efforts on a small collection of stars that may harbor a twin of the Earth.

What Do They Look Like?

Other scientists have tried to use physics, biology, and chemistry to guess what alien life might look like. Isaac Newton, for example, wondered why all the animals he could see around him possessed the same bilateral symmetry–two eyes, two arms, and two legs arranged symmetrically. Was this a fortuitous accident or an act of God?

Today biologists believe that during the "Cambrian explosion," about half a billion years ago, nature experimented with a vast array of shapes and forms for tiny, emerging multicellular creatures. Some had spinal cords shaped like an X, Y, or Z. Some had radial symmetry like a starfish. By accident one had a spinal cord shaped like an I, with bilateral symmetry, and it was the ancestor of most mammals on Earth. So in principle the humanoid shape with bilateral symmetry, the same shape that Hollywood uses to depict aliens in space, does not necessarily have to apply to all intelligent life.

Some biologists believe that the reason that diverse life-forms flourished during the Cambrian explosion is because of an "arms race" between predator and prey. The emergence of the first multicelled organisms that could devour other organisms forced an accelerated evolution of the two, with each one racing to outmaneuver the other. Like the arms race between the former Soviet Union and the United States during the cold war, each side had to hustle to keep ahead of the other.

By examining how life evolved on this planet, one may also make the following speculations about how intelligent life might have evolved on Earth. Scientists have concluded that intelligent life probably requires

1. Some sort of eyesight or sensing mechanism to explore its environment;

2. Some sort of thumb used for grabbing–it could also be a tentacle or claw;
3. Some sort of communication system, such as speech.

These three characteristics are required for sensing our environment and eventually manipulating it–both of which are the hallmarks of intelligence. But beyond these three characteristics, anything goes. Contrary to so many of the aliens shown on TV, an extraterrestrial does not have to resemble a human at all. The child-like, bug-eyed aliens we see on TV and in the movies, in fact, look suspiciously like the 1950s aliens from B-grade movies, which are firmly buried in our unconscious.

(Some anthropologists, however, have added a fourth criteria for intelligent life to explain a curious fact: humans are vastly more intelligent than they have to be to survive in the forest. Our brains can master space travel, the quantum theory, and advanced mathematics–skill sets that are totally unnecessary for hunting and scavenging in the forest. Why this excess brainpower? In nature when we see pairs of animals like the cheetah and the antelope that possess extraordinary skills far beyond those required for survival, we find that there was an arms race between them. Similarly, some scientists believe there is a fourth criteria, a biological "arms race" propelling intelligent humans. Perhaps that arms race was with other members of our own species.)

Think of all the remarkably diverse life-forms on the Earth. If one, for example, could selectively breed octopods for several million years, it is conceivable that they might also become intelligent. (We separated from the apes 6 million years ago, probably because we were not well adapted to the changing environment of Africa. By contrast, the octopus is very well adapted to its life beneath a rock and hence has not evolved for millions of years.) Biochemist Clifford Pickover says that when he gazes at all the "crazy-looking crustaceans, squishy-tentacled jellyfish, grotesque, hermaphroditic worms, and slime molds, I know that God has a sense of humor, and we will see this reflected in other forms in the universe."

Hollywood, however, probably gets it right when it depicts intelligent alien life-forms as carnivores. Not only do meat-eating aliens guarantee bigger box office sales, there is also an element of truth to the depiction. Predators are usually smarter than their prey. Predators have to use cunning to plan, stalk, hide, and ambush prey. Foxes, dogs, tigers, and lions have eyes that are on the front of their face in order to judge distance when they pounce on their prey. With two eyes, they can use 3-D stereo-vision to lock on to their prey. Prey, such as deer and rabbits, on the other hand, just have to know how to run. They have eyes that are on the side of their face in order to scan for predators 360 degrees around them.

In other words, intelligent life in outer space may very well evolve from predators with eyes, or some sensing organ, on the front of their face. They may possess some of the carnivorous, aggressive, and territorial behavior we find in wolves, lions, and humans on Earth. (But since such life-forms would probably be based on entirely different DNA and protein molecules, they would have no interest in eating, or mating with, us.)

We can also use physics to conjecture what their body size might be. Assuming they live on Earth-sized planets and have the same rough density as water, like life-forms on Earth, then huge creatures are probably not possible because of the *scale law,* which states that the laws of physics change drastically as we increase the scale of any object.

MONSTERS AND THE SCALE LAW

If King Kong really existed, for example, he would not be able to terrorize New York City. On the contrary, his legs would break as soon as he took a single step. This is because if you take an ape and increase his size by 10 times, then his weight would go up by the increased volume, or by $10 \times 10 \times 10 = 1{,}000$ times. So he would be 1,000 times heavier. But his strength increases relative to the thickness of his bones and muscles. The cross-sectional area of his bones and muscles goes up by only a square of the distance, that is, by $10 \times 10 = 100$ times. In other words, if King Kong were 10 times bigger, he would be only 100 times stronger, but he would weigh 1,000 times more. Thus the ape's weight increases much more rapidly than its strength as we increase its size. He would be, relatively speaking, 10 times weaker than a normal ape. And that is why his legs would break.

In elementary school I remember my teacher marveling at the strength of an ant, which can lift a leaf many times its weight. My teacher concluded that if an ant were the size of a house, it could pick up that house. But this assumption is incorrect for the same reason that we just saw with King Kong. If an ant were the size of a house, its legs would also break. If you scale up an ant by a factor of 1,000, then it would be 1,000 times weaker than a normal ant, and hence it would collapse of its own weight. (It would also suffocate. An ant breathes through holes in the sides of its body. The area of these holes grows as per the square of the radius, but the volume of the ant increases as per the cube of the radius. Thus an ant 1,000 times bigger than an ordinary ant would have 1,000 times less air than necessary to supply oxygen for its muscles and body tissue. This is also the reason that champion figure skaters and gymnasts tend to be much shorter than average, although they have the same proportions as anyone else. Pound for pound, they have greater proportionate muscle strength than taller people.)

Using the scale law, we can also calculate the rough shape of animals on Earth, and possibly aliens in space. The heat emitted by an animal increases as its surface area increases. Hence increasing its size by 10 increases its heat loss by $10 \times 10 = 100$. But the heat content within its body is proportional to its volume, or $10 \times 10 \times 10 = 1{,}000$. Hence, large animals lose heat more slowly than small animals. (This is the reason that in wintertime our fingers and ears freeze first, since they have the most relative surface area, and why small people get colder faster than large people. It explains why newspapers burn very quickly, because of their large relative surface area, while logs burn very slowly, because of their relatively small surface area.) It also explains why whales in the Arctic are round in shape—because a sphere has the smallest possible surface area per unit mass. And why insects in a warmer environment can afford to be spindly in shape, with a relatively large surface area per unit mass.

In the Disney movie *Honey, I Shrunk the Kids* a family is shrunk down to the size of ants. A rainstorm develops, and in the microworld we see tiny raindrops falling onto puddles. In reality a raindrop as seen by an ant would appear not as a tiny drop but as a huge mound or hemisphere of water. In our world a hemispherical mound of water is unstable and will collapse of its own weight under gravity. But in the microworld, surface tension is relatively large, so a hemispherical mound of water is perfectly stable.

Similarly, in outer space we can estimate the rough surface-to-volume ratio of animals on distant planets using the laws of physics. Using these laws we can theorize that aliens in outer space would likely not be the giants often portrayed in science fiction, but would more closely resemble us in size. (Whales, however, can be much larger in size because of the buoyancy of seawater. This also explains why a beached whale dies–because it is crushed by its own weight.)

The scale law means that the laws of physics change as we go deeper and deeper into the microworld. This explains why the quantum theory appears so bizarre to us, violating simple commonsense notions about our universe. So the scale law rules out the familiar idea of worlds-within-worlds found in science fiction, that is, the idea that inside the atom there could be an entire universe, or that our galaxy could be an atom in a much larger universe. This idea was explored in the movie *Men in Black.* In the final scene of the movie the camera pans away from the Earth, to the planets, the stars, the galaxies, until our entire universe becomes just a single ball in a huge extraterrestrial game played by gigantic aliens.

In reality a galaxy of stars bears no resemblance to an atom; inside the atom the electrons inside their shells are totally different from planets. We know that all the planets are quite different from each other and can orbit at any distance from the mother star. In atoms, however, all the subatomic particles are identical to one another. They cannot orbit at any distance from the nucleus, but only in discrete orbits. (Furthermore, unlike planets, electrons can exhibit bizarre behavior that violates common sense, such as being two places at the same time and having wavelike properties.)

THE PHYSICS OF ADVANCED CIVILIZATIONS

It is also possible to use physics to sketch out the outlines of possible civilizations in space. If we look at the rise of our own civilization over the past 100,000 years, since modern humans emerged in Africa, it can be seen as the story of rising energy consumption. Russian astrophysicist Nikolai Kardashev has conjectured that the stages in the development of extraterrestrial civilizations in the universe could also be ranked by energy consumption. Using the laws of physics, he grouped the possible civilizations into three types:

1. Type I civilizations: those that harvest planetary power, utilizing all the sunlight that strikes their planet. They can, perhaps, harness the power of volcanoes, manipulate the weather, control earthquakes, and build cities on the ocean. All planetary power is within their control.

2. Type II civilizations: those that can utilize the entire power of their sun, making them 10 billion times more powerful than a Type I civilization. The Federation of Planets in *Star Trek* is a Type II civilization. A Type II civilization, in a sense, is immortal; nothing known to science, such as ice ages, meteor impacts, or even supernovae, can destroy it. (In case their mother star is about to explode, these beings can move to another star system, or perhaps even move their home planet.)

3. Type III civilizations: those that can utilize the power of an entire galaxy. They are 10 billion times more powerful than a Type II civilization. The Borg in *Star Trek*, the Empire in *Star Wars*, and the galactic civilization in Asimov's Foundation series correspond to a Type III civilization. They have

> colonized billions of star systems and can exploit the power of the black hole at the center of their galaxy. They freely roam the space lanes of the galaxy.

Kardashev estimated that any civilization growing at a modest rate of a few percent per year in energy consumption will progress rapidly from one type to the next, within a matter of a few thousand years to tens of thousands of years.

As I've discussed in my previous books, our own civilization qualifies a Type 0 civilization (i.e., we use dead plants, oil and coal, to fuel our machines). We utilize only a tiny fraction of the sun's energy that falls on our planet. But already we can see the beginnings of a Type I civilization emerging on the Earth. The Internet is the beginning of a Type I telephone system connecting the entire planet. The beginning of a Type I economy can be seen in the rise of the European Union, which in turn was created to compete with NAFTA. English is already the number one second language on the Earth and the language of science, finance, and business. I imagine it may become the Type I language spoken by virtually everyone. Local cultures and customs will continue to thrive in thousands of varieties on the Earth, but superimposed on this mosaic of peoples will be a planetary culture, perhaps dominated by youth culture and commercialism.

The transition between one civilization and the next is far from guaranteed. The most dangerous transition, for example, may be between a Type 0 and a Type I civilization. A Type 0 civilization is still wracked with the sectarianism, fundamentalism, and racism that typified its rise, and it is not clear whether or not these tribal and religious passions will overwhelm the transition. (Perhaps one reason that we don't see Type I civilizations in the galaxy is because they never made the transition, i.e., they self-destructed. One day, as we visit other star systems, we may find the remains of civilizations that killed themselves in one way or another, e.g., their atmospheres became radioactive or too hot to sustain life.)

By the time a civilization has reached Type III status it has the energy and know-how to travel freely throughout the galaxy and even reach the planet Earth. As in the movie *2001,* such civilizations may well send self-replicating, robotic probes throughout the galaxy searching for intelligent life.

But a Type III civilization would likely not be inclined to visit us or conquer us, as in the movie *Independence Day,* where such a civilization spreads like a plague of locusts, swarming around planets to suck their resources dry. In reality, there are countless dead planets in outer space with vast mineral wealth they could harvest without the nuisance of coping with a restive native population. Their attitude toward us might resemble our own attitude toward an ant hill. Our inclination is not to bend down and offer the ants beads and trinkets, but simply to ignore them.

The main danger ants face is not that humans want to invade them or wipe them out. Instead it is simply that we will pave them over because they are in the way. Remember that the distance between a Type III civilization and our own Type 0 civilization is far more vast than the distance between us and the ants, in terms of energy usage.

UFOs

Some people claim that extraterrestrials have already visited the Earth in the form of UFOs. Scientists usually roll their eyes when they hear about UFOs and dismiss the possibility because the distances between stars are so vast. But regardless of scientists' reactions, persistent reports of UFOs have not diminished over the years.

UFO sightings actually date back to the beginning of recorded history. In the Bible the prophet Ezekiel mentions enigmatically "wheels within wheels in the sky," which some believe is a reference to a UFO. In 1450 BC, during the reign of Pharaoh Thutmose III in Egypt, the Egyptian scribes recorded an incident involving "circles of fire" brighter than the sun, about 5 meters in size, which appeared on several days and finally ascended into the sky. In 91 BC the Roman author Julius Obsequens wrote about "a round object, like a globe, a round or circular shield [that] took its path in the sky." In 1235 General Yoritsume and his army saw strange globes of light dancing in the sky near Kyoto, Japan. In 1561 a large number of objects were seen over Nuremberg, Germany, as if engaged in an aerial battle.

More recently, the U.S. Air Force has conducted large-scale studies of UFO sightings. In 1952 the Air Force began Project Blue Book, which analyzed a total of 12,618 sightings. The report concluded that the vast majority of these sightings could be explained by natural phenomena, conventional aircraft, or hoaxes. Yet about 6 percent were classified as being of unknown origin. But as a result of the Condon Report, which concluded that there was nothing of value in such studies, Project Blue Book was closed in 1969. It was the last known large-scale UFO research project of the U.S. Air Force.

In 2007 the French government released its voluminous file on UFOs to the general public. The report, made available over the Internet by the French National Center for Space Studies, included 1,600 UFO sightings spanning fifty years, including 100,000 pages of eyewitness accounts, films, and audiotapes. The French government stated that 9 percent of these sightings could be fully explained, and that 33 percent have likely explanations, but that they were unable to follow up on the rest.

It is hard, of course, to independently verify these sightings. In fact, most UFO reports, on careful analysis, can be dismissed as the result of the following:

1. *The planet Venus, which is the brightest object in the night sky after the moon.* Because of its enormous distance from the Earth, the planet appears to follow you if you are moving in a car, giving the illusion that it is being piloted, the same way that the moon appears to follow you. We judge distance, in part, by comparing moving objects to their surroundings. Since the moon and Venus are so far away, with nothing to compare them to, they do not move with respect to our surroundings, and hence give us the optical illusion that they are following us.

2. *Swamp gas.* During a temperature inversion over a swampy area, gas will hover over the ground and can become slightly incandescent. Smaller pockets of gas might separate from a larger pocket, giving the impression that scout ships are leaving the "mother ship."

3. *Meteors.* Bright streaks of light can travel across the night sky in a matter of seconds, giving the illusion of a piloted ship. They can also break up, again giving the illusion of scout ships leaving the mother ship.

4. *Atmospheric anomalies.* There are all sorts of lightning storms and unusual atmospheric events that can illuminate the sky in strange ways, giving the illusion of a UFO.

In the twentieth and twenty-first centuries the following phenomena might also generate UFO sightings:

1. *Radar echoes.* Radar waves can bounce off mountains and create echoes, which can be picked up by radar monitors. Such waves even appear to zigzag and fly at enormous velocities on a radar screen, because they are just echoes.

2. *Weather and research balloons.* The military claims, in a controversial report, that the famous rumor of a 1947 alien crash at Roswell, New Mexico, was caused by an errant balloon from Project Mogul, a top-secret project to monitor radiation levels in the atmosphere in case nuclear war broke out.

3. *Aircraft.* Commercial and military aircraft have been known to set off UFO reports. This is particularly true of test flights by advanced experimental aircraft, such as the stealth bomber. (The U.S. military actually encouraged stories of flying saucers in order to deflect attention away from its top-secret projects.)

4. *Deliberate hoaxes.* Some of the most famous pictures that claim to capture flying saucers are actually hoaxes. One

well-known flying saucer, showing windows and landing pods, was actually a modified chicken feeder.

At least 95 percent of the sightings can be dismissed as one of the above. But this still leaves open the question of the remaining few percent of unexplained cases. The most credible cases of UFOs involve (a) multiple sightings by independent, credible eyewitnesses, and (b) evidence from multiple sources, such as eyesight and radar. Such reports are harder to dismiss, since they involve several independent checks. For example, in 1986 there was a sighting of a UFO by JAL-Flight 1628 over Alaska, which was investigated by the FAA. The UFO was seen by the passengers of the JAL flight and was also tracked by ground radar. Similarly, there were mass radar sightings of black triangles over Belgium in 1989–90 that were tracked by NATO radar and jet interceptors. In 1976 there was a sighting over Tehran, Iran, that resulted in multiple systems failures in an F-4 jet interceptor, as recorded in CIA documents.

What is frustrating to scientists is that, of the thousands of recorded sightings, none has produced hard physical evidence that can lead to reproducible results in the laboratory. No alien DNA, alien computer chip, or physical evidence of an alien landing has ever been retrieved.

Assuming for the moment that such UFOs might be real spacecraft rather than illusions, we might ask ourselves what kind of spacecraft they would be. Here are some of the characteristics that have been recorded by observers.

a. They are known to zigzag in midair.
b. They have been known to stop car ignitions and disrupt electrical power as they pass by.
c. They hover silently in the air.

None of these characteristics fit the description of the rockets we have developed on Earth. For example, all known rockets depend on Newton's third law of motion (for every action, there is an equal and opposite reaction); yet the UFOs cited do not seem to have any exhaust whatsoever. And the g-forces created by zigzagging flying saucers would exceed one hundred times the gravitational force on Earth—the g-forces would be enough to flatten any creature on Earth.

Can such UFO characteristics be explained using modern science? In the movies, such as *Earth vs. the Flying Saucers,* it is always assumed that alien beings pilot these craft. More likely, however, if such craft exist they are unmanned (or are manned by a being that is part organic and part mechanical). This would explain how the craft could execute patterns generating g-forces that would normally crush a living being.

A ship that was able to stop car ignitions and move silently in the air suggests a vehicle propelled by magnetism. The problem with magnetic propulsion is that magnets always come with two poles, a north pole and a south pole. If you place a magnet in the Earth's magnetic field, it will simply spin (like a compass needle) rather than rise in the air like a UFO; as the south pole of a magnet moves one way, the north pole moves the opposite way, so the magnet spins and goes nowhere.

One possible solution to this problem would be to use "monopoles," that is, magnets with just one pole, either north or south. Normally if you break a magnet in half you do not get two monopoles. Instead each half of the magnet becomes a magnet by itself, with its own north and south pole; that is, it becomes another dipole. So if you continue to shatter a magnet, you will always find pairs of north and south poles. (This process of breaking a dipole magnet to create smaller dipole magnets continues all the way down to the atomic level, where the atoms themselves are dipoles.)

The problem for scientists is that monopoles have never been seen in the lab. Physicists have tried to photograph the track of a monopole moving through their equipment and have failed (except for a single, highly controversial picture taken at Stanford University in 1982).

Although monopoles have never been conclusively seen experimentally, physicists widely believe that the universe once had an abundance of monopoles at the instant of the big bang. This idea is built into the latest cosmological theories of the big bang. But because the universe inflated rapidly after the big bang, the density of monopoles throughout the universe has been diluted, so we don't see them in the lab today. (In fact, the lack of monopoles today was the key observation that led physicists to propose the inflationary universe idea. So the concept of relic monopoles is well established in physics.)

It is conceivable, therefore, that a space-faring race might be able to harvest these "primordial monopoles" left over from the big bang by throwing out a large magnetic "net" in outer space. Once they have gathered enough monopoles, they can coast through space, using the magnetic field lines found throughout the galaxy or on a planet, without creating exhaust. Because monopoles are the subject of intense interest by many cosmologists, the existence of such a ship is certainly compatible with current thinking in physics.

Lastly, any alien civilization advanced enough to send starships throughout the universe has certainly mastered nanotechnology. This would mean that their starships do not have to be very large; they could be sent by the millions to explore inhabited planets. Desolate moons would perhaps be the best bases for such nanoships. If so, then perhaps our own moon has been visited in the past by a Type III civilization, similar to the scenario depicted in the movie *2001,* which is perhaps the most realistic depiction of an encounter with an extraterrestrial civilization. More than likely, the craft would be unmanned and robotic and placed on the moon. (It may take another century before our technology is advanced enough to scan the entire moon for anomalies in radiation, and is capable of detecting ancient evidence of a previous visitation by nanoships.)

If indeed our moon has been visited in the past or has been the site of a nanotech base, then this might explain why UFOs are not necessarily very large. Some scientists have scoffed at UFOs because they don't fit any of the gigantic propulsion designs being considered by engineers today, such as ramjet fusion engines, huge laser-powered sails, and nuclear pulsed engines, which might be miles across. UFOs can be as small as a jet airplane. But if there is a permanent moon base left over from a previous visitation, then UFOs do not have to be large; they can refuel from their nearby moon base. So sightings may correspond to unmanned reconnaissance ships that originate from the moon base.

Given the rapid advances in SETI and discovering extrasolar planets, contact with extraterrestrial life, assuming it exists in our vicinity, may occur within this century, making such contact a Class I impossibility. If alien civilizations do exist in outer space, the next obvious questions are: Will we ever have the means to reach them? And what about our own distant future, when the sun begins to expand and devour the Earth? Does our destiny really lie in the stars?

9: STARSHIPS

This foolish idea of shooting at the moon is an example of the absurd length to which vicious specialization will carry scientists . . . the proposition appears to be basically impossible.

–A. W. BICKERTON, 1926

The finer part of mankind will, in all likelihood, never perish–they will migrate from sun to sun as they go out. And so there is no end to life, to intellect and the perfection of humanity. Its progress is everlasting.

–KONSTANTIN E. TSIOLKOVSKY, FATHER OF ROCKETRY

One day in the distant future we will have our last nice day on Earth. Eventually, billions of years from now, the sky will be on fire. The sun will swell into a raging inferno that will fill up the entire sky, dwarfing everything in the heavens. As temperatures on Earth soar, the oceans will boil and evaporate, leaving a scorched, parched landscape. The mountains will eventually melt and turn liquid, creating lava flows where vibrant cities once stood.

According to the laws of physics, this grim scenario is inevitable. The Earth will eventually die in flames as it is consumed by the sun. This is a law of physics.

This calamity will take place within the next five billion years. On such a cosmic time scale, the rise and fall of human civilizations are but tiny ripples. One day we must leave the Earth or die. So how will humanity, our descendants, cope when conditions on Earth become intolerable?

Mathematician and philosopher Bertrand Russell once lamented "that no fire, no heroism, no intensity of thought or feeling, can preserve a life beyond the grave; that all the labors of the ages, all the devotion, all the inspiration, all the noonday brightness of human genius, are destined to extinction in the vast death of the solar system; and the whole temple of Man's achievement must inevitably be buried beneath the debris of a universe in ruins . . ."

To me this is one of the most sobering passages in the English language. But Russell wrote this passage in an era when rocket ships were considered impossible. Today the prospect of one day leaving the Earth is not so far-fetched. Carl Sagan once said we should become a "two planet species." Life on Earth is so precious, he said, that we should spread to at least one other inhabitable planet in case of a catastrophe. The Earth moves in the middle of a "cosmic shooting gallery" of asteroids, comets, and other debris drifting near the orbit of the Earth, and a collision with any one of them could result in our demise.

Catastrophes to Come

Poet Robert Frost asked the question whether the Earth will end in fire or ice. Using the laws of physics, we can reasonably predict how the world will end in the event of a natural catastrophe.

On a scale of millennia, one danger to human civilization is the emergence of a new ice age. The last ice age ended 10,000 years ago. When the next one arrives 10,000 to 20,000 years from now most of North America may be covered in half a mile of ice. Human civilization has flourished within the recent tiny interglacial period, when the Earth has been unusually warm, but such a cycle cannot last forever.

Over the course of millions of years, large meteors or comets colliding with Earth could have a devastating impact. The last big celestial impact took place 65 million years ago, when an object about 6 miles across slammed into the Yucatán Peninsula of Mexico, creating a crater about 180 miles in diameter, wiping out the dinosaurs that up until then were the dominant life-form on Earth. Another cosmic collision is likely on that time scale.

Billions of years from now the sun will gradually expand and consume the Earth. In fact, we estimate that the sun will heat up by approximately 10 percent over the next billion years, scorching the Earth. It will completely consume the Earth in 5 billion years, when our sun mutates into a gigantic red star. The Earth will actually be inside the atmosphere of the sun.

Tens of billions of years from now both the sun and the Milky Way galaxy will die. As our sun eventually exhausts its hydrogen/helium fuel, it will shrink into a tiny white dwarf star and gradually cool off until it becomes a hulk of black nuclear waste drifting in the vacuum of space. The Milky Way galaxy will eventually collide with the neighboring Andromeda galaxy, which is much larger than our galaxy. The Milky Way's spiral arms will be torn apart, and our sun could well be flung into deep space. The black holes at the center of the two galaxies will perform a death dance before ultimately colliding and merging.

Given that humanity must one day flee the solar system to the nearby stars to survive, or perish, the question is: how will we get there? The nearest star system, Alpha Centauri, is over 4 light-years away. Conventional chemical propulsion rockets, the workhorses of the current space program, barely reach 40,000 miles per hour. At that speed it would take 70,000 years just to visit the nearest star.

Analyzing the space program today, there is an enormous gap between our pitiful present-day capabilities and the requirements for a true starship that could enable us to begin to explore the universe. Since exploring the moon in the early 1970s, our manned space program has sent astronauts into orbit only about 300 miles above the Earth in the Space Shuttle and International Space Station. By 2010, however, NASA plans to phase out the Space Shuttle to make way for the Orion spacecraft, which will eventually take astronauts back to the moon by the year 2020, after a fifty-year hiatus. The plan is to establish a permanent, manned moon base. A manned mission may be launched to Mars after that.

Obviously a new kind of rocket design must be found if we are ever to reach the stars. Either we must radically increase the thrust of our rockets, or we need to increase the time over which our rockets operate. A large chemical rocket, for example, may have the thrust of several million pounds, but it burns for only a few minutes. By contrast, other rocket designs, such as the ion engine (described in the following paragraphs), may have a feeble thrust but can operate for years in outer space. When it comes to rocketry, the tortoise wins over the hare.

Ion and Plasma Engines

Unlike chemical rockets, ion engines do not produce the sudden, dramatic blast of superhot gases that propel conventional rockets. In fact, their thrust is often measured in ounces. Placed on a tabletop on Earth, they are too feeble to move. But what they lack in thrust they more than make up for in duration, because they can operate for years in the vacuum of outer space.

A typical ion engine looks like the inside of a TV tube. A hot filament is heated by an electric current, which creates a beam of ionized atoms, such as xenon, that is shot out the end of the rocket. Instead of riding on a blast of hot, explosive gas, ion engines ride on a thin but steady flow of ions.

NASA's NSTAR ion thruster was tested in outer space aboard the successful *Deep Space 1* probe, launched in 1998. The ion engine fired for a total of 678 days, setting a new record for ion engines. The European Space Agency has also tested an ion engine on its *Smart 1* probe. The Japanese Hayabusa space probe, which flew past an asteroid, was powered by four xenon ion engines. Although unglamorous, the ion engine will be able to make long-haul missions (that are not urgent) between the planets. In fact, ion engines may one day become the workhorse for interplanetary transport.

A more powerful version of the ion engine is the plasma engine, for example, the VASIMR (variable specific impulse magnetoplasma rocket), which uses a powerful jet of plasma to propel it through space. Designed by astronaut/engineer Franklin Chang-Diaz, it uses radio waves and magnetic fields to heat hydrogen gas to a million degrees centigrade. The superhot plasma is then ejected out the end of the rocket, yielding significant thrust. Prototypes of the engine have already been built on Earth, although none has ever been sent into space. Some engineers hope the plasma engine can be used to power a mission to Mars, significantly reducing the travel time to Mars, down to a few months. Some designs use solar power to energize the plasma in the engine. Other designs use nuclear fission (which raises safety concerns, since it involves putting large amounts of nuclear materials into space on ships that are susceptible to accident).

Neither the ion nor the plasma/VASIMR engine, however, has enough power to take us to the stars. For that, we need an entirely new set of propulsion designs. One serious drawback to designing a starship is the staggering amount of fuel necessary to make a trip to even the nearest star, and the long span of time before the ship reaches its distant destination.

Solar Sails

One proposal that may solve these problems is the solar sail. It exploits the fact that sunlight exerts a very small but steady pressure that is sufficient to propel a huge sail through space. The idea for a solar sail is an old one, dating back to the great astronomer Johannes Kepler in his 1611 treatise *Somnium*.

Although the physics behind a solar sail is simple enough, progress has been spotty in actually creating a solar sail that can be sent into space. In 2004 a Japanese rocket successfully deployed two small prototype solar sails into space. In 2005 the Planetary Society, Cosmos Studios, and the Russian Academy of Sciences launched the *Cosmos 1* space sail from a submarine in the Barents Sea, but the Volna rocket it was being carried on failed, and the sail did not reach orbit. (A previous attempt at a suborbital sail also failed back in 2001.) But in February 2006 a 15-meter solar sail was sent successfully into orbit by the Japanese M-V rocket, although the sail opened incompletely.

Although progress in solar sail technology has been painfully slow, proponents of the solar sail have another idea that might take them to the stars: building a huge battery of lasers on the moon that can fire intense beams of laser light at a solar sail, enabling it to coast to the nearest star. The physics of such an interplanetary solar sail are truly daunting. The sail itself would have to be hundreds of miles across and constructed entirely in outer space. One would have to build thousands of powerful laser beams on the moon, each capable of firing continuously for years to decades. (In one estimate, it would be necessary to fire lasers that have one thousand times the current total power output of the planet Earth.)

On paper a mammoth light sail might be able to travel as fast as half the speed of light. It would take such a solar sail only eight years or so to reach the nearby stars. The advantage of such a propulsion system is that it could use off-the-shelf technology. No new laws of physics would have to be discovered to create such a solar sail. But the main problems are economic and technical. The engineering problems in creating a sail hundreds of miles across, energized by thousands of powerful laser beams placed on the moon, are formidable, requiring a technology that may be a century in the future. (One problem with the interstellar solar sail is coming back. One would have to create a second battery of laser beams on a distant moon to propel the vessel back to Earth. Or perhaps the ship could swing rapidly around a star, using it like a slingshot to get enough speed for the return voyage. Then lasers on the moon would be used to decelerate the sail so it could land on the Earth.)

Ramjet Fusion

My own favorite candidate for getting us to the stars is the ramjet fusion engine. There is an abundance of hydrogen in the universe, so a ramjet engine could scoop hydrogen as it traveled in outer space, essentially giving it an inexhaustible source of rocket fuel. Once the hydrogen was collected it would then be heated to millions of degrees, hot enough so that the hydrogen would fuse, releasing the energy of a thermonuclear reaction.

The ramjet fusion engine was proposed by physicist Robert W. Bussard in 1960 and later popularized by Carl Sagan. Bussard calculated that a ramjet engine weighing about 1,000 tons might theoretically be able to maintain a steady thrust of 1 g of force, that is, comparable to standing on the surface of the Earth. If the ramjet engine could maintain a 1 g acceleration for one year, it would reach 77 percent of the velocity of light, sufficient to make interstellar travel a serious possibility.

The requirements for the ramjet fusion engine are easy to compute. First, we know the average density of hydrogen gas throughout the universe. We also can calculate roughly how much hydrogen gas must be burned in order to attain 1 g accelerations. That calculation, in turn, determines how big the "scoop" must be in order to gather hydrogen gas. With a few reasonable assumptions, one can show that you would need a scoop that is about 160 kilometers in diameter. Although creating a scoop of this size would be prohibitive on Earth, building it in outer space poses fewer problems because of weightlessness.

In principle the ramjet engine could propel itself indefinitely, ultimately reaching distant star systems in the galaxy. Since time slows down inside the rocket, according to Einstein, it might be possible to reach astronomical distances without resorting to putting the crew into suspended animation. After accelerating at 1 g for eleven years, according to clocks inside the starship, the spacecraft would reach the Pleiades star cluster, which is 400 light-years away. In twenty-three years it would reach the Andromeda galaxy, which is 2 million light-years from Earth. In theory, the spacecraft might be able to reach the limit of the visible universe within the lifetime of a crew member (although billions of years might have passed on the Earth).

One key uncertainty is the fusion reaction. The ITER fusion reactor, scheduled to be built in the south of France, combines two rare forms of hydrogen (deuterium and tritium) in order to extract energy.

In outer space, however, the most abundant form of hydrogen consists of a single proton surrounded by an electron. The ramjet fusion engine would therefore have to exploit the proton-proton fusion reaction. Although the deuterium/tritium fusion process has been studied for decades by physicists, the proton-proton fusion process is less well understood, is more difficult to achieve, and yields far less power. So mastering the more difficult proton-proton reaction will be a technical challenge in the coming decades. (Some engineers, in addition, have questioned whether the ramjet engine could overcome drag effects as it approaches the speed of light.)

Until the physics and economics of proton-proton fusion are worked out, it is difficult to make accurate estimates as to the ramjet's feasibility. But this design is on the short list of possible candidates for any mission contemplated to the stars.

Nuclear Electric Rocket

In 1956 the U.S. Atomic Energy Commission (AEC) began to look at nuclear rockets seriously under Project Rover. In theory, a nuclear fission reactor would be used to heat up gases like hydrogen to extreme temperatures, and then these gases would be ejected out one end of the rocket, creating thrust.

Because of the danger of an explosion in the Earth's atmosphere involving toxic nuclear fuel, early versions of nuclear rocket engines were placed horizontally on railroad tracks, where the performance of the rocket could be carefully monitored. The first nuclear rocket engine to be tested under Project Rover was the Kiwi 1 in 1959 (aptly named after the Australian flightless bird). In the 1960s NASA joined with the AEC to create the Nuclear Engine for Rocket Vehicle Applications (NERVA), which was the first nuclear rocket to be tested vertically, rather than horizontally. In 1968 this nuclear rocket was test-fired in a downward position.

The results of this research have been mixed. The rockets were very complicated and often misfired. The intense vibrations of the nuclear engine often cracked the fuel bundles, causing the ship to break apart. Corrosion due to burning hydrogen at high temperatures was also a persistent problem. The nuclear rocket program was finally closed in 1972.

(These atomic rockets had yet another problem: the danger of a runaway nuclear reaction, as in a small atomic bomb. Although commercial nuclear power plants today run on diluted nuclear fuel and cannot explode like a Hiroshima bomb, these atomic rockets, in order to create maximum thrust, operated on highly enriched uranium and hence could explode in a chain reaction, creating a tiny nuclear detonation. When the nuclear rocket program was about to be retired, scientists decided to perform one last test. They decided to blow up a rocket, like a small atomic bomb. They removed the control rods [which keep the nuclear reaction in check]. The reactor went supercritical and blew up in a fiery ball of flames. This spectacular demise of the nuclear rocket program was even captured on film. The Russians were not pleased. They considered this stunt to be a violation of the Limited Test Ban Treaty, which banned above-ground detonations of nuclear bombs.)

Over the years the military has periodically revisited the nuclear rocket. One secret project was called the Timberwind nuclear rocket; it was part of the military's Star Wars project in the 1980s. (It was abandoned after details of its existence were released by the Federation of American Scientists.)

The main concern about the nuclear fission rocket is safety. Even fifty years into the space age, chemical booster rockets undergo catastrophic failure about 1 percent of the time. (The two failures of the *Challenger* and *Columbia* Space Shuttles, tragically killing fourteen astronauts, further confirmed this failure rate.)

Nonetheless, in the past few years NASA has resumed research on the nuclear rocket for the first time since the NERVA program of the 1960s. In 2003 NASA christened a new project, Prometheus, named for the Greek god who gave fire to humanity. In 2005 Prometheus was funded at $430 million, although that funding was significantly cut to $100 million in 2006. The project's future is unclear.

Nuclear Pulsed Rockets

Another distant possibility is to use a series of mini-nuclear bombs to propel a starship. In Project Orion, mini-atomic bombs were to be ejected out the back of the rocket in sequence, so that the spacecraft would "ride" on the shock waves created by these mini-hydrogen bombs. On paper such a design could take a spacecraft close to the speed of light. Originally conceived in 1947 by Stanislaw Ulam, who helped to design the first hydrogen bombs, the idea was further developed by Ted Taylor (one of the chief designers of nuclear warheads for the U.S. military) and physicist Freeman Dyson of the Institute for Advanced Study at Princeton.

In the late 1950s and 1960s elaborate calculations were made for this interstellar rocket. It was estimated that such a starship could make it to Pluto and back within a year, with a top cruising velocity of 10 percent the speed of light. But even at that speed it would take about forty-four years to reach the nearest star. Scientists have speculated that a space ark powered by such a rocket would have to cruise for centuries, with a multigenerational crew whose offspring would be born and spend all their lives on the space ark, in order that their descendants could reach the nearby stars.

In 1959 General Atomics issued a report estimating the size of an Orion spacecraft. The largest version, called the super Orion, would weigh 8 million tons, have a diameter of 400 meters, and be energized by over 1,000 hydrogen bombs.

But one major problem with the project was the possibility of contamination via nuclear fallout during launch. Dyson estimated that the nuclear fallout from each launch could cause fatal cancers in ten people. In addition, the electromagnetic pulse (EMP) for such a launch would be so great that it could cause massive short circuits in neighboring electrical systems.

The signing of the Limited Test Ban Treaty in 1963 sounded the death knell of the project. Eventually the main driving force pushing the project, nuclear bomb designer Ted Taylor, gave up. (He once confided to me that he finally became disillusioned with the project when he realized that the physics behind mini-nuclear bombs could also be used by terrorists to create portable nuclear bombs. Although the project was canceled because it was deemed too dangerous, its namesake lives on in the Orion spacecraft, which NASA has chosen to replace the Space Shuttle in 2010.)

The concept of a nuclear-fired rocket was briefly resurrected by the British Interplanetary Society from 1973 to 1978, with Project Daedalus, a preliminary study to see if an unmanned starship could be built that could reach the Barnard's Star, 5.9 light-years from Earth. (Barnard's Star was chosen because it was conjectured that it might have a planet. Since then astronomers Jill Tarter and Margaret Turnbull have compiled a list of 17,129 nearby stars that could have planets supporting life. The most promising candidate is Epsilon Indi A, 11.8 light-years away.)

The rocket ship planned for Project Daedalus was so huge that it would have had to be constructed in outer space. It would weigh 54,000 tons, nearly all of its weight in rocket fuel, and could attain 7.1 percent of the speed of light with a payload of 450 tons. Unlike Project Orion, which used tiny fission bombs, Project Daedalus would use mini-hydrogen bombs with a deuterium/helium-3 mixture ignited by electron beams. Because of the formidable technical problems facing it, as well as concerns over its nuclear propulsion system, Project Daedalus was also shelved indefinitely.

Specific Impulse and Engine Efficiency

Engineers sometimes speak of "specific impulse," which enables us to rank the efficiency of various engine designs. "Specific impulse" is defined as the change in momentum per unit mass of propellant. Hence the more efficient the engine, the less fuel is necessary to boost a rocket into space. Momentum, in turn, is the product of the force acting over a period of time. Chemical rockets, although they have very large thrust, operate for only a few minutes, and hence have a very low specific impulse. Ion engines, because they can operate for years, can have high specific impulse with very low thrust.

Specific impulse is measured in seconds. A typical chemical rocket might have a specific impulse of 400–500 seconds. The specific impulse of the Space Shuttle engine is 453 seconds. (The highest specific impulse ever achieved for a chemical rocket was 542 seconds, using a propellant mixture of hydrogen, lithium, and fluorine.) The thruster for the *Smart 1* ion engine had a specific impulse of 1,640 seconds. And the nuclear rocket attained specific impulses of 850 seconds.

The maximum possible specific impulse would be a rocket that could attain the speed of light. It would have a specific impulse of about 30 million. Following is a table showing the specific impulses of different kinds of rocket engines.

TYPE OF ROCKET ENGINE	SPECIFIC IMPULSE
Solid fuel rocket	250
Liquid fuel rocket	450
Ion engine	3,000
VASIMR plasma engine	1,000 to 30,000
Nuclear fission rocket	800 to 1,000
Nuclear fusion rocket	2,500 to 200,000
Nuclear pulsed rocket	10,000 to 1 million
Antimatter rocket	1 million to 10 million

(In principle, laser sails and ram-jet engines, because they contain no rocket propellant at all, have infinite specific impulse, although they have problems of their own.)

Space Elevators

One severe objection to many of these rocket designs is that they are so mammoth and heavy that they could never be built on the Earth. That is why some scientists have proposed building them in outer space, where weightlessness would make it possible for astronauts to lift impossibly heavy objects with ease. But critics today point out the prohibitive costs of assembly in outer space. The International Space

Station, for example, will require upwards of one hundred launches of shuttle missions for complete assembly and costs have escalated to $100 billion. It is the most expensive scientific project in history. Building an interstellar space sail or ramjet scoop in outer space would cost many times that amount.

But as science fiction writer Robert Heinlein was fond of saying, if you can make it to 160 kilometers above the Earth, you are halfway to anywhere in the solar system. That is because the first 160 kilometers of any launch, when the rocket is struggling to escape the Earth's gravity, cost by far the most. After that a rocket ship can almost coast to Pluto and beyond.

One way to reduce costs drastically in the future would be to develop a space elevator. The idea of climbing a rope to heaven is an old one, for example, as in the fairy tale "Jack and the Beanstalk," but it might become a reality if the rope could be sent far into space. Then the centrifugal force of the Earth's rotation would be enough to nullify the force of gravity, so the rope would never fall. The rope would magically rise vertically into the air and disappear into the clouds. (Think of a ball spinning on a string. The ball seems to defy gravity, because the centrifugal force pushes it away from the center of rotation. In the same way, a very long rope would be suspended in air because of the spinning of the Earth.) Nothing would be needed to hold up the rope except the spin of the Earth. A person could theoretically climb the rope and ascend into space. We sometimes give the undergraduates taking physics courses at City University of New York the problem of calculating the tension on such a rope. It is easy to show that the tension on the rope would be enough to snap even a steel cable, which is why building a space elevator has long been considered to be impossible.

The first scientist to seriously study the space elevator was Russian visionary scientist Konstantin Tsiolkovsky. In 1895, inspired by the Eiffel Tower, he envisioned a tower that would ascend into space, connecting the Earth to a "celestial castle" in space. It would be built bottom-up, starting on Earth, and engineers would slowly extend the space elevator to the heavens.

In 1957 Russian scientist Yuri Artsutanov proposed a new solution, that the space elevator be built in reverse order, top-down, starting from outer space. He envisioned a satellite in a geostationary orbit 36,000 miles in space, where it would appear to be stationary, and from which one would drop a cable down to Earth. Then the cable would be anchored to the ground. But the tether for a space elevator would have to be able to withstand roughly 60–100 gigapascals (gpa) of tension. Steel breaks at about 2 gpa, making the idea beyond reach.

The idea of a space elevator reached a much wider audience with the publication of Arthur C. Clarke's 1979 novel, *The Fountains of Paradise,* and Robert Heinlein's 1982 novel, *Friday.* But without any further progress, the idea languished.

The equation changed significantly when carbon nanotubes were developed by chemists. Interest was suddenly sparked by the work of Sumio Iijima of Nippon Electric in 1991 (although evidence for carbon nanotubes actually dates back to the 1950s, a fact that was ignored at the time). Remarkably, nanotubes are much stronger than steel cables, but also much lighter. In fact, they exceed the strength necessary to maintain a space elevator. Scientists believe a carbon nanotube fiber could withstand 120 gpa of pressure, which is comfortably above the breaking point. This discovery has rekindled attempts to create a space elevator.

In 1999 a NASA study gave serious consideration to the space elevator, envisioning a ribbon, about 1 meter wide and about 47,000 kilometers long, capable of transporting about 15 tons of payload into Earth's orbit. Such a space elevator could change the economics of space travel overnight. The cost could be reduced by a factor of ten thousand, an astonishing, revolutionary change.

Currently it costs $10,000 or more to send a pound of material into orbit around the Earth (roughly the cost, ounce for ounce, of gold). Each Space Shuttle mission, for example, costs up to $700 million. A space elevator could reduce the cost to as little as $1 per pound. Such a radical reduction in the cost of the space program could revolutionize the way we view space travel. With a simple push of an elevator button, one could in principle take an elevator ride into outer space for the price of a plane ticket.

But formidable practical hurdles have to be solved before we build a space elevator on which we can levitate our way into heaven. At present pure carbon nanotube fibers created in the lab are no more than 15 millimeters long. To create a space elevator, one would have to create cables of carbon nanotubes that are thousands of miles long. Although from a scientific point of view this is just a technical problem, it is a stubborn and difficult problem that must be solved if we are to create a space elevator. Yet, within a few decades, many scientists believe that we should be able to master the technology of creating long cables of carbon nanotubes.

Second, microscopic impurities in the carbon nanotubes could make a long cable problematic. Nicola Pugno of the Polytechnic of Turin, Italy, estimates that if a carbon nanotube has even one atom misaligned, its strength could be reduced by 30 percent. Overall, atomic-scale defects could reduce the strength of the nanotube cable by as much as 70 percent, taking it below the minimum gigapascals of strength necessary to support a space elevator.

To spur entrepreneurial interest in the space elevator, NASA is funding two separate prizes. (The prizes are modeled on the $10 million Ansari X-prize, which successfully spurred enterprising inventors to create commercial rockets capable of taking passengers to the very edge of space. The X-prize was won by Spaceship One in 2004.) The prizes NASA is offering are called the Beam Power Challenge and the Tether Challenge. In the Beam Power Challenge, teams have to send a mechanical device weighing at least 25 kilograms up a tether (suspended from a crane) at the speed of 1 meter per second for a distance of 50 meters. This may sound easy, but the catch is that the device cannot use fuel, batteries, or an electrical cord. Instead, the robot device must be powered by solar arrays, solar reflectors, lasers, or microwaves–energy sources that are more suitable for use in outer space.

In the Tether Challenge, teams must produce 2-meter-long tethers that cannot weigh more than 2 grams and must carry 50 percent more weight than the best tether of the previous year. The challenge is intended to stimulate research in developing lightweight materials strong enough to be strung 100,000 kilometers in space. There are prizes worth $150,000, $40,000, and $10,000. (To highlight the difficulty of mastering this challenge, in 2005, the first year of the competition, no one won a prize.)

Although a successful space elevator could revolutionize the space program, such machines have their own sets of hazards. For example, the trajectory of near-Earth satellites constantly shifts as they orbit the Earth (this is because the Earth rotates beneath them). This means that these satellites would eventually collide with the space elevator at 18,000 miles per hour, sufficient to rupture the tether. To prevent such a catastrophe, in the future either satellites will have to be designed to include small rockets so that they can maneuver around the space elevator, or the tether of the elevator might have to be equipped with small rockets to evade passing satellites.

Also, collisions with micrometeorites are a problem, since the space elevator is far above the atmosphere of the Earth, and our atmosphere usually protects us from meteors. Since micrometeor collisions are unpredictable, the space elevator must be built with added shielding and perhaps even fail-safe redundancy systems. Problems could also emerge from the effects of turbulent weather patterns on the Earth, such as hurricanes, tidal waves, and storms.

The Slingshot Effect

Another novel means of hurling an object near the speed of light is to use the "slingshot" effect. When sending space probes to the outer planets, NASA sometimes whips them around a neighboring planet, so they use the slingshot effect to boost their velocity. NASA saves on valuable rocket fuel in this way. That's how the *Voyager* spacecraft was able to reach Neptune, which lies near the very edge of the solar system.

Princeton physicist Freeman Dyson proposed that in the far future we might find two neutron stars that are revolving around each other at great speed. By traveling extremely close to one of these neutron stars, we could whip around it and then be hurled into space at speeds approaching a third the speed of light. In effect, we would be using gravity to give us an additional boost to nearly the speed of light. On paper this just might work.

Others have proposed that we whip around our own sun in order to accelerate to near the speed of light. This method, in fact, was used in *Star Trek IV: The Voyage Home,* when the crew of the *Enterprise* hijacked a Klingon ship and then sped close to the Sun in order to break the light barrier and go back in time. In the movie *When Worlds Collide,* when Earth is threatened by a collision with an asteroid, scientists flee the Earth by creating a gigantic roller coaster. A rocket ship descends the roller coaster, gaining great velocity, and then whips around the bottom of the roller coaster to blast off into space.

In fact, however, neither of these methods of using gravity to boost our way into space will work. (Because of the conservation of energy, in going down a roller coaster and coming back up, we wind up with the *same* velocity as that with which we started, so there is no gain in energy whatsoever. Likewise, by whipping around the stationary sun, we wind up with the same velocity as that with which we originally started.) The reason Dyson's method of using two neutron stars might work is because the neutron stars are revolving so fast. A spacecraft using the slingshot effect gains its energy from the motion of a planet or star. If they are stationary, there is no slingshot effect at all.

Although Dyson's proposal could work, it does not help today's Earth-bound scientists, because we would need a starship just to visit rotating neutron stars.

Rail Guns to the Heavens

Yet another ingenious method for flinging objects into space at fantastic velocities is the rail gun, which Arthur C. Clarke and others have featured in their science fiction tales, and which is also being seriously examined as part of the Star Wars missile shield.

Instead of using rocket fuel or gunpowder to boost a projectile to high velocity, a rail gun uses the power of electromagnetism.

In its simplest form, a rail gun consists of two parallel wires or rails, with a projectile that straddles both wires, forming a U-shaped configuration. Even Michael Faraday knew that a current of electricity will experience a force when placed in a magnetic field. (This, in fact, is the basis of all electrical motors.) By sending millions of amperes of electrical power down these wires and through the projectile, a huge magnetic field is created around the rails. This magnetic field then propels the projectile down the rails at enormous velocities.

Rail guns have successfully fired metal objects at enormous velocities over extremely short distances. Remarkably, in theory, a simple rail gun should be able to fire a metal projectile at 18,000 miles per hour, so that it would go into orbit around the Earth. In principle, NASA's entire rocket fleet could be replaced by rail guns that could blast payloads into orbit from the Earth.

The rail gun enjoys a significant advantage over chemical rockets and guns. In a rifle the ultimate velocity at which expanding gases can push a bullet is limited by the speed of shock waves. Although Jules Verne used gunpowder to blast astronauts to the moon in his classic tale *From the Earth to the Moon*, one can compute that the ultimate velocity that one can attain with gunpowder is only a fraction of the velocity necessary to send someone to the moon. Rail guns, however, are not limited by the speed of shock waves.

But there are problems with the rail gun. It accelerates objects so fast that they usually flatten upon impact with the air. Payloads have been severely deformed in the process of being fired out of the barrel of a rail gun because when the projectile hits the air it's like hitting a wall of bricks. In addition, the huge acceleration of the payload along the rails is enough to deform them. The tracks have to be replaced regularly because of the damage caused by the projectile. Furthermore, the g-forces on an astronaut would be enough to kill him, easily crushing all the bones in his body.

One proposal is to install a rail gun on the moon. Outside the Earth's atmosphere, a rail gun's projectile could speed effortlessly through the vacuum of outer space. But even then the enormous accelerations generated by a rail gun might damage the payload. Rail guns in some sense are the opposite of laser sails, which build up their ultimate speed gently over a long period of time. Rail guns are limited because they pack so much energy into such a small space.

Rail guns that can fire objects to nearby stars would be quite expensive. In one proposal the rail gun would be built in outer space, extending two-thirds of the distance from Earth to the sun . It would store solar energy from the sun and then abruptly discharge that energy into the rail gun, sending a 10-ton payload at one-third the speed of light, with an acceleration of 5000 g's. Not surprisingly, only the sturdiest robotic payloads would be able to survive such huge accelerations.

The Dangers of Space Travel

Of course, space travel is no Sunday picnic. Enormous dangers await manned flights traveling to Mars, or beyond. Life on Earth has been sheltered for millions of years: The planet's ozone layer protects the Earth from ultraviolet rays, its magnetic field protects against solar flares and cosmic rays, and its thick atmosphere protects against meteors that burn up on entry. We take for granted the mild temperatures and air pressures found on the Earth. But in deep space, we must face the reality that most of the universe is in turmoil, with lethal radiation belts and swarms of deadly meteors.

The first problem to solve in extended space travel is that of weightlessness. Long-term studies of weightlessness by the Russians have shown that the body loses precious minerals and chemicals in space much faster than expected. Even with a rigorous exercise program, after a year on the space station, the bones and muscles of Russian cosmonauts are so atrophied that they can barely crawl like babies when they first return to Earth. Muscle atrophy, deterioration of the skeletal system, lower production of red blood cells, lower immune response, and a reduced functioning of the cardiovascular system seem to be the inevitable consequences of prolonged weightlessness in space.

Missions to Mars, which may take several months to a year, will push the very limits of the endurance of our astronauts. For long-term missions to the nearby stars, this problem could be fatal. The starships of the future may have to spin, creating an artificial gravity via centrifugal forces in order to sustain human life. This adjustment would greatly increase the cost and complexity of future spaceships.

Second, the presence of micrometeorites in space traveling at many tens of thousands of miles per hour may require that spaceships be equipped with extra shielding. Close examination of the hull of the Space Shuttle has revealed evidence of several tiny but potentially deadly impacts from tiny meteorites. In the future, spaceships may have to contain a special doubly reinforced chamber for the crew.

Radiation levels in deep space are much higher than previously thought. During the eleven-year sunspot cycle, for example, solar flares can send enormous quantities of deadly plasma racing toward Earth. In the past, this phenomenon has forced the astronauts on the space station to seek special protection against the potentially lethal barrage of subatomic particles. Space walks during such solar eruptions would be fatal. (Even taking a simple transatlantic trip from L.A. to New York, for example, exposes us to about a millirem of radiation per hour of flight. Over the course of our trip we are exposed to almost a dental X-ray of radiation.) In deep space, where the atmosphere and magnetic field of the Earth no longer protect us, radiation exposure could be a serious problem.

Suspended Animation

One consistent criticism of the rocket designs I have presented so far is that even if we could build such starships, it would take decades to centuries to reach nearby stars. Such a mission would need to involve a multigenerational crew whose descendants would arrive at the final destination.

One solution, proposed in such movies as *Alien* and *Planet of the Apes,* is for space travelers to undergo suspended animation; that is, their body temperature would be carefully lowered until bodily functions almost cease. Animals that hibernate do this every year during the winter. Certain fish and frogs can be frozen solid in a block of ice and yet thaw out when the temperature rises.

Biologists who have studied this curious phenomenon believe that these animals have the ability to create a natural "antifreeze" that lowers the freezing point of water. This natural antifreeze consists of certain proteins in fish, and glucose in frogs. By flooding their blood with these proteins, fish can survive in the Arctic at about –2°C. Frogs have evolved the ability to maintain high glucose levels, thereby preventing the formation of ice crystals. Although their bodies might be frozen solid on the outside, they are not frozen on the inside, allowing their bodily organs to continue to operate, albeit at a reduced rate.

There are problems with adapting this ability to mammals, however. When human tissue is frozen, ice crystals begin to form inside the cells. As these ice crystals grow, they can penetrate and destroy cell walls. (Celebrities who want to have their heads and bodies frozen in liquid nitrogen after death may want to think twice.)

Nevertheless, there has been recent progress in limited suspended animation in mammals that do not naturally hibernate, such as mice and dogs. In 2005 scientists at the University of Pittsburgh were able to bring dogs back to life after their blood had been drained and replaced by a special ice-cold solution. Clinically dead for three hours, the dogs were brought back to life after their hearts were restarted. (Although most of the dogs were healthy after this procedure, a few suffered some brain damage.)

That same year scientists were able to place mice in a chamber containing hydrogen sulfide and successfully reduce their body temperature to 13°C for six hours. The metabolism rate of the mice dropped by a factor of ten. In 2006 doctors at Massachusetts General

Hospital in Boston placed pigs and mice in a state of suspended animation using hydrogen sulfide.

In the future such procedures may be lifesaving for people involved in severe accidents or who suffer heart attacks during which every second counts. Suspended animation might allow doctors to "freeze time" until patients can be treated. But it could be decades or more before such techniques can be applied to human astronauts, who may need to be in suspended animation for centuries.

NANOSHIPS

There are several other ways in which we might be able to reach the stars via more advanced, unproven technologies that border on science fiction. One promising proposal is to use unmanned probes based on nanotechnology. Throughout this discussion I have assumed that starships need to be monstrous devices consuming vast amounts of energy, capable of taking a large crew of human beings to the stars, similar to the starship *Enterprise* on *Star Trek.*

But a more likely avenue might be initially to send miniature unmanned probes to the distant stars at near the speed of light. As we mentioned earlier, in the future, with nanotechnology, it should be possible to create tiny spacecraft that exploit the power of atomic and molecular-sized machines. For example, ions, because they are so light, can easily be accelerated to near the speed of light with ordinary voltages found in the laboratory. Instead of requiring huge booster rockets, they might be sent into space at near the speed of light using powerful electromagnetic fields. This means that if a nanobot were ionized and placed within an electric field, it could effortlessly be boosted to near light speed. The nanobot would then coast its way to the stars, since there is no friction in space. In this way, many of the problems plaguing large starships are immediately solved. Unmanned intelligent nanobot spaceships might be able to reach nearby star systems at a mere fraction of the cost of building and launching a huge starship carrying a human crew.

Such nanoships could be used to reach nearby stars or, as Gerald Nordley, a retired Air Force astronautical engineer, has suggested, to push against a solar sail in order to propel it through space. Nordley says, "With a constellation of pinhead-sized spacecraft flying in formation and communicating with themselves, you could practically push them with a flashlight."

But there are challenges with nano starships. They might be deflected by passing electric and magnetic fields in outer space. To counteract these forces, one would need to accelerate the nanoships to very high voltages on the Earth so they would not be easily deflected. Second, we might have to send a swarm of millions of these nanobot starships to guarantee that a handful would actually make it to their destination. Sending a swarm of starships to explore the nearest stars might seem extravagant, but such starships would be cheap and could be mass-produced by the billions, so that only a tiny fraction of them would have to reach their target.

What might these nanoships look like? Dan Goldin, former head of NASA, envisioned a fleet of "Coke-can sized" spacecraft. Others have talked about starships the size of needles. The Pentagon has been looking into the possibility of developing "smart dust," dust-sized particles that have tiny sensors inside that can be sprayed over a battlefield to give commanders real-time information. In the future it is conceivable that "smart dust" might be sent to the nearby stars.

Dust-sized nanobots would have their circuitry made by the same etching techniques used in the semiconductor industry, which can create components as small as 30 nm, or roughly 150 atoms across. These nanobots could be launched from the moon by rail guns or even by particle accelerators, which regularly send subatomic particles to near light speed. These devices would be so cheap to make that millions of them could be launched into space.

Once they reached a nearby star system, the nanobots could land on a desolate moon. Because of the moon's low gravity, a nanobot would be able to land and take off with ease. And with a stable environment such as a moon would provide, it would make an ideal base of operations. The nanobot could build a nanofactory, using the minerals found on the moon, to create a powerful radio station that could beam information back to Earth. Or the nanofactory could be designed to create millions of copies of itself to explore the solar system and venture off to other nearby stars, repeating the process. Because these ships would be robotic, there would be no need for a return voyage home once they had radioed back their information.

The nanobot I've just described is sometimes called a von Neumann probe, named after the famed mathematician John von Neumann, who worked out the mathematics of self-replicating Turing machines. In principle, such self-replicating nanobot spaceships might be able to explore the entire galaxy, not just the nearby stars. Eventually there could be a sphere of trillions of these robots, multiplying exponentially as it grows in size, expanding at nearly the speed of light. The nanobots inside this expanding sphere could colonize the entire galaxy within a few hundred thousand years.

One electrical engineer who takes the idea of nanoships very seriously is Brian Gilchrist of the University of Michigan. He recently received a $500,000 grant from NASA's Institute for Advanced Concepts to explore the idea of building nanoships with engines no bigger than a bacterium. He envisions using the same etching technology used in the semiconductor industry to create a fleet of several million nanoships that will propel themselves by ejecting tiny nanoparticles that are only tens of nanometers across. These nanoparticles would be energized by passing through an electric field, just as in an ion engine. Since each nanoparticle weighs thousands of times more than an ion, the engines would pack much more thrust than a typical ion engine. Thus the nanoship engines would have the same advantages as an ion engine, except they would have much more thrust. Gilchrist has already begun etching some of the parts for these nanoships. So far he can pack 10,000 individual thrusters on a single silicon chip that measures 1 centimeter across. Initially he envisions sending his fleet of nanoships throughout the solar system to test their efficiency. But eventually these nanoships might be part of the first fleet to reach the stars.

Gilchrist's proposal is one of several futuristic proposals being considered by NASA. After several decades of inactivity, NASA has recently given some serious thought to various proposals for interstellar travel–proposals that range from the credible to the fantastic. Since the early 1990s NASA has hosted the annual Advanced Space Propulsion Research Workshop, during which these technologies have been picked apart by teams of serious engineers and physicists. Even more ambitious is the Breakthrough Propulsion Physics program, which has explored the mysterious world of quantum physics in relation to interstellar travel. Although there is no consensus, much of their activity has focused on the front-runners: the laser sail and various versions of fusion rockets.

Given the slow but steady advances in spaceship design, it is reasonable to assume that the first unmanned probe of some sort might be sent to the nearby stars perhaps later in this century or early in the next century, making it a Class I impossibility.

But perhaps the most powerful design for a starship involves the use of antimatter. Although it sounds like science fiction, antimatter has already been created on the Earth, and may one day provide the most promising design yet for a workable manned starship.

10: ANTIMATTER AND ANTI-UNIVERSES

The most exciting phrase to hear in science, the one that heralds new discoveries, is not "Eureka" (I found it!) but "That's funny . . ."

–ISAAC ASIMOV

If the man doesn't believe as we do, we say he is a crank, and that settles it. I mean, it does nowadays, because now we can't burn him.

–MARK TWAIN

You can recognize a pioneer by the arrows in his back.

–BEVERLY RUBIK

In Dan Brown's book *Angels and Demons,* the bestselling predecessor to *The Da Vinci Code,* a small band of extremists, the Illuminati, have hatched a plot to blow up the Vatican using an antimatter bomb, stolen from CERN, the nuclear laboratory outside Geneva. The conspirators know that when matter and antimatter touch each other the result is a monumental explosion, many times more powerful than a hydrogen bomb. Although an antimatter bomb is pure fiction, antimatter is very real.

An atomic bomb, for all its awesome power, is only about 1 percent efficient. Only a tiny fraction of the uranium is turned into energy. But if an antimatter bomb could be constructed, it would convert 100 percent of its mass into energy, making it far more efficient than a nuclear bomb. (More precisely, about 50 percent of the matter in an antimatter bomb would be turned into usable explosive energy; the rest would be carried away in the form of undetectable particles called neutrinos.)

Antimatter has long been the focus of intense speculation. Although an antimatter bomb does not exist, physicists have been able to use their powerful atom smashers to create minute quantities of antimatter for study.

Producing Anti-atoms and Anti-chemistry

At the beginning of the twentieth century, physicists realized that the atom consisted of charged subatomic particles with electrons (with a negative charge) circulating around a tiny nucleus (with a positive charge). The nucleus, in turn, consisted of protons (which carried the positive charge) and neutrons (which were electrically neutral).

So it came as quite a shock in the 1930s when physicists realized that for every particle there is a twin, an antiparticle, but with an opposite charge. The first antiparticle to be discovered was the antielectron (called the positron), which has a positive charge. The positron is identical to the electron in every way, except that it carries the opposite charge. It was first discovered in photographs of cosmic rays taken in a cloud chamber. (Positron tracks are quite easy to see in a cloud chamber. When placed in a powerful magnetic field, they bend in the opposite direction from ordinary electrons. In fact, I photographed such antimatter tracks while I was in high school.)

In 1955 the particle accelerator at the University of California at Berkeley, the Bevatron, produced the first antiproton. As expected, it is identical to the proton except that it has a negative charge. This means that, in principle, one can create anti-atoms (with positrons circulating around antiprotons). In fact, anti-elements, anti-chemistry, anti-people, anti-Earths, and even anti-universes are theoretically possible.

At present the giant particle accelerators at CERN and the Fermilab outside Chicago have been able to create minute quantities of antihydrogen. (This is done by blasting a beam of high-energy protons into a target using particle accelerators, thereby creating a shower of subatomic debris. Powerful magnets separate out the antiprotons, which are slowed down to very low velocities and then are exposed to the antielectrons that are naturally emitted from sodium-22. When the antielectrons orbit around the antiprotons, they create antihydrogen, since hydrogen is made up of one proton and one electron.) In a pure vacuum, these anti-atoms might live forever. But because of impurities and collisions with the wall, these anti-atoms eventually strike ordinary atoms and they are annihilated, releasing energy.

In 1995 CERN made history when it announced that it had created nine antihydrogen atoms. Fermilab soon followed suit by producing one hundred atoms of antihydrogen. In principle, there is nothing to prevent us from creating higher anti-elements as well, except for the staggering cost. Producing even a few ounces of anti-atoms would bankrupt any nation. The current rate of production of antimatter is between one-billionth to ten-billionths of a gram per year. The yield might increase by a factor of three by the year 2020. The economics of antimatter are very poor. In 2004 it cost CERN $20 million to produce several trillionths of a gram of antimatter. At that rate, producing a single gram of antimatter would cost $100 quadrillion and the antimatter factory would need to run continuously for 100 billion years! This makes antimatter the most precious substance in the world.

"If we could assemble all the anti-matter we've ever made at CERN and annihilate it with matter," reads a statement from CERN, "we would have enough energy to light a single electric light bulb for a few minutes."

Handling antimatter poses extraordinary problems, since any contact between matter and antimatter is explosive. Putting antimatter in an ordinary container would be suicide. When the antimatter touched the walls, it would explode. So how does one handle antimatter if it is so volatile? One way would be first to ionize the antimatter into a gas of ions, and then to safely confine it in a "magnetic bottle." The magnetic field would prevent the antimatter from touching the walls of the chamber.

To build an antimatter engine, a steady stream of antimatter would need to be fed into a reaction chamber, where it would be carefully combined with ordinary matter, creating a controlled explosion, similar to the explosion created by chemical rockets. The ions created by this explosion would then be shot out one end of the antimatter rocket, creating propulsion. Because of the antimatter engine's efficiency in converting matter into energy, in theory it is one of the most appealing engine designs for future starships. In the *Star Trek* series, antimatter is the source of the *Enterprise*'s energy; its engines are energized by the controlled collision of matter and antimatter.

An Antimatter Rocket

One of the main proponents of the antimatter rocket is physicist Gerald Smith of Pennsylvania State University. He believes that in the short term as little as 4 milligrams of positrons would be sufficient to take an antimatter rocket to Mars in just several weeks. He notes that the energy packed into antimatter is about a billion times greater than the energy packed into ordinary rocket fuel.

The first step in creating this fuel would be to create beams of antiprotons, via a particle accelerator, and then store them in a "Penning trap," which Smith is constructing. When built, the Penning trap would weigh 220 pounds (much of it being liquid nitrogen and liquid helium) and would store about a trillion antiprotons in a magnetic field. (At very low temperatures, the wavelength of the antiprotons is several times longer than the wavelength of the atoms in the container walls, so the antiprotons would mainly reflect off the walls without annihilating themselves.) He states that this Penning trap should be able to store the antiprotons for about five days (until they finally are annihilated when mixed with ordinary atoms). His Penning trap should be able to store about a billionth of a gram of antiprotons. His goal is to create a Penning trap that can store up to a microgram of antiprotons.

Although antimatter is the most precious substance on Earth, its cost keeps dropping dramatically every year (a gram would cost about $62.5 trillion at today's prices). A new particle injector being built at Fermilab outside Chicago should be able to increase the production of antimatter by a factor of ten, from 1.5 to 15 nanograms per year, which should drive down prices. However, Harold Gerrish of NASA believes that with further improvements the cost could realistically go down to $5,000 per microgram. Dr. Steven Howe, of Synergistics Technologies in Los Alamos, New Mexico, states, "Our goal is to remove antimatter from the far-out realm of science fiction into the commercially exploitable realm for transportation and medical applications."

So far, particle accelerators that can produce antiprotons are not specifically designed to do so, so they are quite inefficient. Such particle accelerators are designed primarily to be research tools, not factories for antimatter. That is why Smith envisions building a new particle accelerator that will be specifically designed to produce copious quantities of antiprotons to drive down the cost.

If prices for antimatter can be lowered even further by technical improvements and mass production, Smith envisions a time when the antimatter rocket could become a workhorse for interplanetary and possibly interstellar travel. Until then, however, antimatter rockets will remain on the drawing boards.

NATURALLY OCCURRING ANTIMATTER

If antimatter is so difficult to create on Earth, might one find antimatter more easily in outer space? Unfortunately, searches for antimatter in the universe have turned up very little, which is rather surprising to physicists. The fact that our universe is made up mainly of matter, rather than antimatter, is difficult to explain. One might naïvely have assumed that at the beginning of the universe, there were equal, symmetrical quantities of matter and antimatter. So the lack of antimatter is puzzling.

The most likely solution was first proposed by Andrei Sakharov, the man who designed the hydrogen bomb for the Soviet Union in the 1950s. Sakharov theorized that at the beginning of the universe there was a slight asymmetry in the amount of matter and antimatter in the big bang. This tiny symmetry breaking is called "CP violation." This phenomenon is currently the center of much vigorous research. In effect, Sakharov theorized that all the atoms in the universe today are left over from a near perfect cancellation between matter and antimatter; the big bang caused a cosmic cancellation between the two. The tiny leftover matter created a residue that forms the visible universe of today. All the atoms in our bodies are leftovers from this titanic collision of matter and antimatter.

This theory leaves open the possibility that small amounts of antimatter may occur naturally. If so, discovering that source would drastically reduce the cost of producing antimatter for use in antimatter engines. In principle, deposits of naturally occurring antimatter should be easy to detect. When an electron and an antielectron meet, they annihilate into gamma rays at an energy of 1.02 million electron volts or more. Thus by scanning the universe for gamma rays at this energy, one could find the "fingerprint" for naturally occurring antimatter.

In fact, "fountains" of antimatter have been found in the Milky Way galaxy, not far from the galactic center, by Dr. William Purcell of Northwestern University. Apparently a stream of antimatter exists that creates this characteristic gamma radiation at 1.02 million electron volts as it collides with ordinary hydrogen gas. If this plume of antimatter exists naturally, then it might be possible that other pockets of antimatter exist in the universe that were not destroyed in the big bang.

To look for naturally occurring antimatter more systematically, the PAMELA (Payload for Antimatter-Matter Exploration and Light-Nuclei Astrophysics) satellite was launched into orbit in 2006. It is a collaborative effort between Russia, Italy, Germany, and Sweden, designed to search for pockets of antimatter. Previous missions searching for antimatter were carried out using high-altitude balloons and the Space

Shuttle, so the data was collected for no more than a week or so. PAMELA, by contrast, will stay in orbit for at least three years. "It is the best detector ever constructed and we will use it for a long period," declares team member Piergiorgio Picozza of the University of Rome.

PAMELA is designed to detect cosmic rays from ordinary sources, such as supernovae, but also from unusual ones, such as stars made entirely of antimatter. Specifically, PAMELA will look for the signature of anti-helium, which might be produced in the interiors of anti-stars. Although most physicists today believe that the big bang resulted in a near perfect cancellation between matter and antimatter, as Sakharov believed, PAMELA is based on a different assumption–that whole regions of antimatter universe did not undergo that cancellation and hence exist today in the form of anti-stars.

If antimatter exists in minute quantities in deep space, then it might be possible to "harvest" some of that antimatter to use to propel a starship. NASA's Institute for Advanced Concepts takes the idea of harvesting antimatter in space seriously enough that it recently funded a pilot program to study this concept. "Basically, what you want to do is generate a net, just like you're fishing," says Gerald Jackson of Hbar Technologies, one of the organizations spearheading the project.

The antimatter harvester is based on three concentric spheres, each made out of a lattice wire network. The outermost sphere would be 16 kilometers across and would be positively charged, so that it would repel any protons, which are positively charged, but attract antiprotons, which are negatively charged. The antiprotons would be collected by the outer sphere, then slow down as they passed through the second sphere and would finally stop when they reached the innermost sphere, which would be 100 meters across. The antiprotons would then be captured in a magnetic bottle and combined with antielectrons to make antihydrogen.

Jackson estimates that controlled matter-antimatter reactions inside a spacecraft could fuel a solar sail to Pluto using just 30 milligrams of antimatter. Seventeen grams of antimatter, says Jackson, would be enough to fuel a starship to Alpha Centauri. Jackson claims that there might be 80 grams of antimatter between the orbits of Venus and Mars that might be harvested by the space probe. Given the complexities and cost of launching this huge antimatter collector, however, it probably won't be realized until the end of this century, or beyond.

Some scientists have dreamed about harvesting antimatter from a meteor floating in outer space. (The *Flash Gordon* comic strip once featured a rogue antimatter meteor drifting in space, which could create a terrifying explosion if it came in contact with any planet.)

If naturally occurring antimatter is not found in space, we will have to wait decades or even centuries before we can produce significantly large quantities of antimatter on the Earth. But assuming that the technical problems of producing antimatter can be solved, this leaves open the possibility that one day antimatter rockets may take us to the stars.

Given what we know of antimatter today, and the foreseeable evolution of this technology, I would classify an antimatter rocket ship as a Class I impossibility.

FOUNDER OF ANTIMATTER

What is antimatter? It seems strange that nature would double the number of subatomic particles in the universe for no good reason. Nature is usually quite sparing, but now that we know about antimatter, nature seems to be supremely redundant and wasteful. And if antimatter exists, can anti-universes also exist?

To answer these questions, one has to investigate the origin of antimatter itself. The discovery of antimatter actually dates back to 1928, with the pioneering work of Paul Dirac, one of the most brilliant physicists of the twentieth century. He held the Lucasian Chair at Cambridge University, the same chair held by Newton, and the chair currently held by Stephen Hawking. Dirac, born in 1902, was a tall, wiry man who was in his early twenties when the quantum revolution broke open in 1925. Although he was studying electrical engineering at that time, he was suddenly swept up in the tidal wave of interest unleashed by the quantum theory.

The quantum theory was based on the idea that particles like electrons could be described not as pointlike particles but as a wave of some sort, described by Schrödinger's celebrated wave equation. (The wave represents the probability of finding the particle at that point.)

But Dirac realized that there was a defect with Schrödinger's equation. It described only electrons moving at low velocities. At higher velocities, the equation failed because it did not obey the laws of objects moving at high velocities, that is, the laws of relativity found by Albert Einstein.

To the young Dirac, the challenge was to reformulate the Schrödinger equation to accommodate the theory of relativity. In 1928 Dirac proposed a radical modification of the Schrödinger equation that fully obeyed Einstein's relativity theory. The world of physics was stunned. Dirac found his famous relativistic equation for the electron purely by manipulating higher mathematical objects, called spinors. A mathematical curiosity was suddenly becoming a centerpiece for the entire universe. (Unlike many physicists before him, who insisted that great breakthroughs in physics be firmly grounded in experimental data, Dirac took the opposite strategy. To him pure mathematics, if it was beautiful enough, was the sure guide to great breakthroughs. He wrote, "It is more important to have beauty in one's equations than to have them fit experiments . . . It seems that if one is working from the point of view of getting beauty in one's equations, and if one has a really sound insight, one is on a sure line of progress.")

In developing his new equation for the electron, Dirac realized that Einstein's celebrated equation, $E = mc^2$, was not quite right. Although it is splattered over Madison Avenue ads, children's T-shirts, cartoons, and even the costumes of superheroes, Einstein's equation is only partially correct. The correct equation is actually $E = \pm mc^2$. (This minus sign arises because we have to take the square root of a certain quantity. Taking the square root of a quantity always introduces a plus or minus ambiguity.)

But physicists abhor negative energy. There is an axiom of physics that states that objects always tend to the state of lowest energy (this is the reason that water always seeks the lowest level, sea level). Since matter always drops down to its lowest energy state, the prospect of negative energy was potentially disastrous. It meant that all electrons would eventually tumble down to infinite negative energy, hence Dirac's theory would be unstable. So Dirac invented the concept of the "Dirac sea." He envisioned that all negative energy states were already filled up, and hence an electron could not tumble down into negative energy. Hence the universe was stable. Also a gamma ray might occasionally collide with an electron sitting in a negative energy state and kick it up into a state of positive energy. We would then see the gamma ray turn into an electron and a "hole" develop in the Dirac sea. This hole would act like a bubble in the vacuum; that is, it would have a positive charge and the same mass as the original electron. In other words, the hole would behave like an antielectron. So in this picture antimatter consists of "bubbles" in the Dirac sea.

Just a few years after Dirac made this astounding prediction, Carl Anderson actually discovered the antielectron (for which Dirac won the Nobel Prize in 1933).

In other words, antimatter exists because the Dirac equation has two types of solutions, one for matter, and one for antimatter. (And this in turn is the outcome of special relativity.)

Not only did the Dirac equation predict the existence of antimatter; it also predicted the "spin" of the electron. Subatomic particles can spin, much like a spinning top. The spin of the electron, in turn, is crucial to understanding the flow of electrons in transistors and semiconductors, which form the basis of modern electronics.

Stephen Hawking regrets that Dirac did not patent his equation. He writes, "Dirac would have made a fortune if he had patented the Dirac equation. He would have had a royalty on every television, Walkman, video game and computer."

Today Dirac's celebrated equation is etched in the stone of Westminster Abbey, not far from the tomb of Isaac Newton. In the entire world, it is perhaps the only equation given this distinctive honor.

Dirac and Newton

Historians of science seeking to understand the origins of how Dirac came up with his revolutionary equation and the concept of antimatter have often compared him to Newton. Strangely, Newton and Dirac share a number of similarities. Both were in their twenties when they did their seminal work at Cambridge University, both were masters of mathematics, and both shared another stark characteristic: a total lack of social skills, to the point of pathology. Both were notorious for their inability to engage in small talk and simple social graces. Painfully shy, Dirac would never say anything unless asked directly, and then he would reply "yes," or "no," or "I don't know."

Dirac was also extremely modest and detested publicity. When he was awarded the Nobel Prize in Physics, he seriously considered turning it down because of the notoriety and trouble that it would generate. But when it was pointed out to him that rejecting the Nobel Prize would generate even more publicity he decided to accept it.

Volumes have been written about Newton's peculiar personality, with hypotheses ranging from mercury poisoning to mental illness. But recently a new theory has been proposed by Cambridge psychologist Simon Baron-Cohen that might explain both Newton's and Dirac's strange personalities. Baron-Cohen claims that they both probably suffered from Asperger's syndrome, which is akin to autism, like the idiot savant in the movie *Rain Man.* Individuals suffering from Asperger's are notoriously reticent, socially awkward, and sometimes blessed with enormous calculational ability, but unlike autistic individuals they are functional in society and can hold productive jobs. If this theory is true, then perhaps the miraculous calculational power of Newton and Dirac came at a price, being socially apart from the rest of humanity.

Antigravity and Anti-universes

Using Dirac's theory, we can now answer a host of questions: What is the antimatter counterpart of gravity? Do anti-universes exist?

As we discussed, antiparticles have the opposite charge of ordinary matter. But particles that have no charge at all (such as the photon, a particle of light, or the graviton, which is a particle of gravity) can be their own antiparticle. We see that gravitation is its own antimatter; in other words, gravity and antigravity are the same thing. Hence antimatter should fall down under gravity, not up. (This is universally believed by physicists, but it has actually never been demonstrated in the laboratory.)

Dirac's theory also answers the deep questions: Why does nature allow for antimatter? Does that mean anti-universes exist?

In some science fiction tales, the protagonist discovers a new Earth-like planet in outer space. In fact, the new planet seems identical to Earth in every way, except everything is made of antimatter. We have antimatter twins on this planet, with anti-children, who live in anti-cities. Since the laws of anti-chemistry are the same as the laws of chemistry, except charges are reversed, people living in such a world would never know they were made of antimatter. (Physicists call this the charge-reversed or C-reversed universe, since all charges are reversed in this anti-universe, but everything else remains the same.)

In other science fiction stories scientists discover a twin of the Earth in outer space, except that it is a Looking Glass universe, where everything is left-right reversed. Everyone's heart is on the right side and most people are left-handed. They live out their lives never knowing that they live in a left-right reversed Looking Glass universe. (Physicists call such a Looking Glass universe a parity-reversed or P-reversed universe.)

Can such antimatter and parity-reversed universes really exist? Physicists take questions about twin universes very seriously, since Newton's and Einstein's equations remain the same when we simply flip the charges on all our subatomic particles or reverse the left-right orientation. Hence, C-reversed and P-reversed universes are in principle possible.

Nobel laureate Richard Feynman posed an interesting question about these universes. Suppose one day we make radio contact with aliens on a distant planet but cannot see them. Can we explain to them the difference between "left" and "right" by radio? he asked. If the laws of physics allow for a P-reversed universe, then it should be impossible to convey these concepts.

Certain things, he reasoned, are easy to communicate, such as the shape of our bodies and the number of our fingers, arms, and legs. We can even explain to the aliens the laws of chemistry and biology. But if we try to explain to them the concept of "left" and "right" (or "clockwise" and "counterclockwise"), we would fail each time. We would never be able to explain to them that our heart is on the left side of our body, in which direction the Earth rotates, or the way a DNA molecule spirals.

So it came as a shock when C. N. Yang and T. D. Lee, both at Columbia University at the time, disproved this cherished theorem. By examining the nature of subatomic particles they showed that the Looking Glass, P-reversed universe cannot exist. One physicist, learning of this revolutionary result, said, "God must have made a mistake." For this earthshaking result, called the "overthrow of parity," Yang and Lee won the Nobel Prize in Physics in 1957.

To Feynman, this conclusion meant that if you are talking to aliens on a radio, it is possible to set up an experiment that could enable you to tell the difference between left- and right-handed universes by radio alone. (For example, electrons emitted from radioactive cobalt-60 do not spin in equal numbers in a clockwise or counterclockwise fashion, but actually spin in a preferred direction, thereby breaking parity.)

Feynman then envisioned that a historic meeting finally takes place between the aliens and humanity. We tell the aliens to stick out their right hand when we first meet, and we will shake hands. If the aliens actually stick out their right hand, then we know that we have successfully communicated to them the concept of "left-right" and "clockwise-counterclockwise."

But Feynman then raised an unsettling thought. What happens if the aliens stick out their left hand instead? This means that we have made a fatal mistake, that we have failed to communicate the concept of "left" and "right." Worse, it means that the alien is actually made of antimatter, and that he performed all the experiments backward, and hence got "left" and "right" mixed up. It means when we shake hands, we will explode!

That was our understanding until the 1960s. It was impossible to tell the difference between our universe and a universe in which everything was made of antimatter and was parity-reversed. If you flipped *both* the parity and the charge, the resulting universe would obey the laws of physics. Parity by itself was overthrown, but charge and parity was still a good symmetry of the universe. So a CP-reversed universe was still possible.

This meant that if we were talking to aliens on the phone, we could not tell the difference between an ordinary universe and one that was both parity- and charge-reversed (i.e., left and right are interchanged, and all matter is turned into antimatter).

Then in 1964 physicists received a second shock: the CP-reversed universe cannot exist. By analyzing the properties of subatomic particles, it is still possible to tell the difference between left-right, clockwise-counterclockwise if you are talking by radio to another CP-reversed universe. For this result, James Cronin and Val Fitch won the Nobel Prize in 1980.

(Although many physicists were upset when the CP-reversed universe was shown to be inconsistent with the laws of physics, in hindsight the discovery was a good thing, as we discussed earlier. If the CP-reversed universe were possible, then the original big bang would have involved precisely the same amount of matter and antimatter, and hence 100 percent annihilation would have taken place, and our atoms would not have been possible! The fact that we exist as a leftover from the annihilation of unequal amounts of matter and antimatter is proof of CP violation.)

Are any reversed anti-universes possible? The answer is yes. Even if parity-reversed and charge-reversed universes are not possible, an anti-universe is still possible, but it would be a strange one. If we reversed the charges, the parity, *and the march of time*, then the resulting universe would obey all the laws of physics. The CPT-reversed universe is allowed.

Time reversal is a bizarre symmetry. In a T-reversed universe, fried eggs jump off the dinner plate, reform on the frying pan, and then jump back into the egg, sealing the cracks. Corpses rise from the dead, get younger, turn into babies, and then jump into their mother's womb.

Common sense tells us that the T-reversed universe is not possible. But the mathematical equations of subatomic particles tell us otherwise. Newton's laws run perfectly well backward or forward. Imagine videotaping a billiard game. Each collision of the balls obeys Newton's laws of motion; running such a videotape would make for a bizarre game, but it is allowed by the laws of Newton.

In the quantum theory things are more complicated. T-reversal by itself violates the laws of quantum mechanics, but the full CPT-reversed universe is allowed. This means that a universe in which left and right are reversed, matter turns into antimatter, and time runs backward is a fully acceptable universe obeying the laws of physics!

(Ironically, we cannot communicate with such a CPT-reversed world. If time runs backward on their planet, it means that everything we tell them by radio will be part of their future, so they would forget everything we told them as soon as we spoke to them. So even though the CPT-reversed universe is allowed under the laws of physics, we cannot talk to any CPT-reversed alien by radio.)

In summary, antimatter engines may give us a realistic possibility for fueling a starship in the distant future, if enough antimatter could be made on Earth, or found in outer space. There is a slight imbalance between matter and antimatter because of CP violation, and this in turn may mean that pockets of antimatter still exist and can be harvested.

But because of the technical difficulties involved in antimatter engines, it may take a century or more to develop this technology, making it a Class I impossibility.

But let's tackle another question: Will faster-than-light starships be possible thousands of years in the future? Are there loopholes to Einstein's famous dictum that "nothing can go faster than light"? The answer, surprisingly, is yes.

Part II

CLASS II IMPOSSIBILITIES

11: FASTER THAN LIGHT

It's quite conceivable that [life] will eventually spread through the galaxy and beyond. So life may not forever be an unimportant trace contaminant of the universe, even though it now is. In fact, I find it a rather appealing view.

—ASTRONOMER ROYAL SIR MARTIN REES

It is impossible to travel faster than the speed of light, and certainly not desirable, as one's hat keeps blowing off.

—WOODY ALLEN

In *Star Wars,* as the *Millennium Falcon* blasts off the desert planet Tatooine, carrying our heroes Luke Skywalker and Han Solo, the ship encounters a squadron of menacing Imperial battleships orbiting the planet. The Empire's battleships fire a punishing barrage of laser blasts at our heroes' ship that steadily break through its force fields. The *Millennium Falcon* is outgunned. Buckling under this withering laser fire, Han Solo yells that their only hope is to make the jump into "hyperspace." In the nick of time the hyperdrive engines spring to life. All the stars around them suddenly implode toward the center of their view screen in converging, blinding streaks of light. A hole opens up, which the *Millennium Falcon* blasts through, reaching hyperspace and freedom.

Science fiction? Undoubtedly. But could it be based on scientific fact? Perhaps. Faster-than-light travel has always been a staple of science fiction, but recently physicists have given serious thought to this possibility.

According to Einstein, the speed of light is the ultimate speed limit in the universe. Even our most powerful atom smashers, which can create energies found only at the center of exploding stars or the big bang itself, cannot hurl subatomic particles at a rate faster than the speed of light. Apparently the speed of light is the ultimate traffic cop in the universe. If so, any hope of our reaching the distant galaxies seems to be dashed.

Or maybe not . . .

EINSTEIN THE FAILURE

In 1902 it was far from obvious that the young physicist Albert Einstein would be hailed as the greatest physicist since Isaac Newton. In fact, that year represented the lowest point in his life. A newly minted Ph.D. student, he was rejected for a teaching job by every university he applied to. (He later found out that his professor Heinrich Weber had written horrible letters of recommendation for him, perhaps in revenge for Einstein's having cut so many of his classes.) Furthermore, Einstein's mother was violently opposed to his girlfriend, Mileva Maric, who was carrying his child. Their first daughter, Lieserl, would be born illegitimate. Young Albert was also a failure at the odd jobs he took. Even his lowly tutoring job abruptly ended when he was fired. In his depressing letters he contemplated becoming a salesman to earn a living. He even wrote to his family that perhaps it would have been better had he never been born, since he was such a burden to his family and lacked any prospects for success in life. When his father died, he felt ashamed that his father had died thinking that his son was a total failure.

Yet later that year Einstein's luck would turn. A friend arranged for him to get a job as a clerk in the Swiss Patent Office. From that lowly position Einstein would launch the greatest revolution in modern history. He would quickly analyze the patents on his desk and then spend hours contemplating problems in physics that had puzzled him since he was a child.

What was the secret of his genius? Perhaps one clue to his genius was his ability to think in terms of physical pictures (e.g., moving trains, accelerating clocks, stretched fabrics) rather than pure mathematics. Einstein once said that unless a theory can be explained to a child, the theory was probably useless; that is, the essence of a theory has to be captured by a physical picture. So many physicists get lost in a thicket of mathematics that lead nowhere. But like Newton before him, Einstein was obsessed by the physical picture; the mathematics would come later. For Newton the physical picture was the falling apple and the moon. Were the forces that made an apple fall identical to the forces that guided the moon in its orbit? When Newton decided that the answer was yes, he created a mathematical architecture for the universe that suddenly unveiled the greatest secret of the heavens, the motion of celestial bodies themselves.

Einstein and Relativity

Albert Einstein proposed his celebrated special theory of relativity in 1905. At the heart of his theory was a picture that even children can understand. His theory was the culmination of a dream he had had since the age of sixteen, when he asked the fateful question: what happens if you outrace a light beam? As a youth, he knew that Newtonian mechanics described the motion of objects on the Earth and in the heavens, and that Maxwell's theory described light. These were the two pillars of physics.

The essence of Einstein's genius was that he recognized that these two pillars were in contradiction. One of them must fall.

According to Newton, you could always outrace a light beam, since there was nothing special about the speed of light. This meant that the light beam must remain stationary as you raced alongside. But as a youth Einstein realized that no one had ever seen a light wave that was totally stationary, that is, like a frozen wave. Hence Newton's theory did not make sense.

Finally, as a college student in Zurich studying Maxwell's theory, Einstein found the answer. He discovered something that even Maxwell did not know: that the speed of light was a constant, no matter how fast you moved. If you raced toward or away from a light beam, it still traveled at the same velocity, but this trait violates common sense. Einstein had found the answer to his childhood question: you can never race alongside a light beam, since it always moves away from you at a constant speed, no matter how fast you move.

But Newtonian mechanics was a tightly constrained system: like pulling on a loose thread, the entire theory could unravel if you made the smallest change in its assumptions. In Newton's theory the passage of time was uniform throughout the universe. One second on the Earth was identical to one second on Venus or Mars. Similarly, meter sticks placed on the Earth had the same length as meter sticks on Pluto. But if the speed of light was always constant no matter how fast you moved, there would need to be a major shakeup in our understanding of space and time. Profound distortions of space and time would have to occur to preserve the constancy of the speed of light.

According to Einstein, if you were in a speeding rocket ship, the passage of time inside that rocket would have to slow down with respect to someone on Earth. Time beats at different rates, depending on how fast you move. Furthermore, the space within that rocket ship would get compressed, so that meter sticks could change in length, depending on your speed. And the mass of the rocket would increase as well. If we were to peer into the rocket with our telescopes, we would see clocks inside the rocket running slowly, people moving in slow motion, and the people would appear flattened.

In fact, if the rocket were traveling at the speed of light, time would apparently stop inside the rocket, the rocket would be compressed to nothing, and the mass of the rocket would be infinite. Since none of these observations make any sense, Einstein stated that nothing can break the light barrier. (Because an object gets heavier the faster it moves, this means that the energy motion is being converted to mass. The precise amount of energy that turns into mass is easy to calculate, and we arrive at the celebrated equation $E = mc^2$ in just a few lines.)

Since Einstein derived his famous equation, literally millions of experiments have confirmed his revolutionary ideas. For example, the GPS system, which can locate your position on the Earth to within a few feet, would fail unless one added in corrections due to relativity. (Since the military depends on the GPS system, even Pentagon generals have to be briefed by physicists concerning Einstein's theory of relativity.) The clocks on the GPS actually change as they speed above the Earth, as Einstein predicted.

The most graphic illustration of this concept is found in atom smashers, in which scientists accelerate particles to nearly the speed of light. At the gigantic CERN accelerator, the Large Hadron Collider, outside Geneva, Switzerland, protons are accelerated to trillions of electron volts, and they move very close to the speed of light.

To a rocket scientist, the light barrier is not much of a problem yet, since rockets can barely travel beyond a few tens of thousands of miles per hour. But within a century or two, when rocket scientists seriously contemplate sending probes to the nearest star (located over 4 light-years from Earth), the light barrier could gradually become a problem.

Loopholes in Einstein's Theory

Over the decades, physicists have tried to find loopholes in Einstein's famous dictum. Some loopholes have been found, but most are not very useful. For example, if one sweeps a flashlight across the heavens, in principle the image of the light beam can exceed the speed of light. In a few seconds, the image of the flashlight moves from one point on the horizon to the opposite point, over a distance that can stretch over hundreds of light-years. But this is of no importance, since no information can be transmitted faster than light in this fashion. The image of the light beam has exceeded the speed of light, but the image carries no energy or information.

Similarly, if we have a pair of scissors, the point at which the blades cross each other moves faster the farther you are from the joining point. If we imagine scissors that are a light-year long, then by closing the blades the crossing point can travel faster than light. (Again, this is not important since the crossing point carries no energy or information.)

Likewise, as I mentioned in Chapter Four, the EPR experiment enables one to send information at speeds faster than the speed of light. (In this experiment, we recall, two electrons are vibrating in unison and then are sent speeding in opposite directions. Because these electrons are coherent, information can be sent between them at speeds faster than the speed of light, but this information is random and hence is useless. EPR machines, hence, cannot be used to send probes to the distant stars.)

To a physicist, the most important loophole came from Einstein himself, who created the general theory of relativity in 1915, a theory that is more powerful than the special theory of relativity. The seeds of general relativity were planted when Einstein considered a children's merry-go-round. As we saw earlier, objects shrink as they approach the speed of light. The faster you move, the more you are squeezed. But in a spinning disk, the outer circumference moves faster than the center. (The center, in fact, is almost stationary.) This means that a ruler stick placed on the rim must shrink, while a ruler placed at the center remains nearly the same, so the surface of the merry-go-round is no longer flat, but is curved. Thus acceleration has the effect of curving space and time on the merry-go-round.

In the general theory of relativity, space-time is a fabric that can stretch and shrink. Under certain circumstances the fabric may stretch faster than the speed of light. Think of the big bang, for example, when the universe was born in a cosmic explosion 13.7 billion years ago. One can calculate that the universe originally expanded faster than the speed of light. (This action does not violate special relativity, since it was empty space–the space between stars–that was expanding, not the stars themselves. Expanding space does not carry any information.)

The important point is that special relativity applies only locally, that is, in your nearby vicinity. In your local neighborhood (e.g., the solar system), special relativity holds, as we confirm with our space probes. But globally (e.g., on cosmological scales involving the universe) we must use general relativity instead. In general relativity, space-time becomes a fabric, and this fabric can stretch faster than light. It can also allow for "holes in space" in which one can take shortcuts through space and time.

Given these caveats, perhaps one way to travel faster than light is to invoke general relativity. There are two ways in which this might be done.

1. *Stretching space.* If you were to stretch the space behind you and contact the space in front of you, then you would have the illusion of having moved faster than light. In fact, you would not have moved at all. But since space has been deformed, it means you can reach the distant stars in a twinkling of an eye.

2. *Ripping space.* In 1935 Einstein introduced the concept of a wormhole. Imagine the Looking Glass of Alice,, a magical device that connects the countryside of Oxford to Wonderland. The wormhole is a device that can connect two universes. When we were in grade school, we learned that the shortest distance between two points is a straight line. But this is not necessarily true, because if we curled a sheet of paper until two points touched, then we would see that the shortest distance between two points is actually a wormhole.

As physicist Matt Visser of Washington University says, "The relativity community has started to think about what would be necessary to take something like warp drive or wormholes out of the realm of science fiction."

Sir Martin Rees, Royal Astronomer of Great Britain, even says,

"Wormholes, extra dimensions, and quantum computers open up speculative scenarios that could transform our entire universe eventually into a 'living cosmos.' "

The Alcubierre Drive and Negative Energy

The best example of stretching space is the Alcubierre drive, proposed by physicist Miguel Alcubierre in 1994 using Einstein's theory of gravity. It is quite similar to the propulsion system seen in *Star Trek*. The pilot of such a starship would be seated inside a bubble (called a "warp bubble") in which everything seemed to appear normal, even as the spacecraft broke the light barrier. In fact, the pilot would think that he was at rest. Yet outside the warp bubble extreme distortions of space-time would occur as the space in front of the warp bubble was compressed. There would be no time dilation, so time would pass normally inside the warp bubble.

Alcubierre admits that *Star Trek* may have had a role to play in his finding this solution. "People in *Star Trek* kept talking about warp drive, the concept that you're warping space," he says. "We already had a theory about how space can or cannot be distorted, and that is the general theory of relativity. I thought there should be a way of using these concepts to see how a warp drive would work." This is probably the first time that a TV show helped to inspire a solution to one of Einstein's equations.

Alcubierre speculates that a journey in his proposed starship would resemble a journey taken on the *Millennium Falcon* in *Star Wars*. "My guess is they would probably see something very similar to that. In front of the ship, the stars would become long lines, streaks. In back, they wouldn't see anything–just black–because the light of the stars couldn't move fast enough to catch up with them," he says.

The key to the Alcubierre drive is the energy necessary to propel the spacecraft forward at faster-than-light velocities. Normally physicists begin with a positive amount of energy in order to propel a starship, which always travels slower than the speed of light. To move beyond this strategy so as to be able to travel faster than the speed of light one would need to change the fuel. A straightforward calculation shows that you would need "negative mass" or "negative energy," perhaps the most exotic entities in the universe, if they exist. Traditionally, physicists have dismissed negative energy and negative mass as science fiction. But we now see that they are indispensable for faster-than-light travel, and they might actually exist.

Scientists have looked for negative matter in nature, but so far without success. (Antimatter and negative matter are two entirely different things. The first exists and has positive energy, but a reversed charge. Negative matter has not yet been proven to exist.) Negative matter would be quite peculiar, because it would be lighter than nothing. In fact, it would float. If negative matter existed in the early universe, it would have drifted into outer space. Unlike meteors that come crashing down onto planets, drawn by a planet's gravity, negative matter would shun planets. It would be repelled, not attracted, by large bodies such as stars and planets. Hence, although negative matter might exist, we expect to find it only in deep space, certainly not on Earth.

One proposal to find negative matter in outer space involves using the phenomenon called "Einstein lenses." When light travels around a star or galaxy its path is bent by its gravity, according to general relativity. In 1912 (even before Einstein fully developed general relativity) he predicted that a galaxy might be able to act like the lens of a telescope. Light from a distant object moving around a nearby galaxy would converge as it passed around the galaxy, like a lens, forming a characteristic ring pattern when the light finally reached the Earth. These phenomena are now called "Einstein rings." In 1979 the first of these Einstein lenses was observed in outer space. Since then, Einstein lenses have become an indispensable tool for astronomers. (For example, it was once thought that it would be impossible to locate "dark matter" in outer space. [Dark matter is a mysterious substance that is invisible but has weight. It surrounds the galaxies and is perhaps ten times as plentiful as ordinary visible matter in the universe.] But NASA scientists have been able to construct maps of dark matter since dark matter bends light as the light passes through, in the same way that glass bends light.)

Therefore it should be possible to use Einstein lenses to search for negative matter and wormholes in outer space. They should bend light in a peculiar way, which should be visible with the Hubble Space Telescope. So far, Einstein lenses have not detected the image of negative matter or wormholes in outer space, but the search is continuing. If one day the Hubble Space Telescope detects the presence of negative matter or a wormhole via Einstein lenses, it could set off a shock wave in physics.

Negative energy is different from negative matter in that it actually exists, but only in minute quantities. In 1933 Hendrik Casimir made a bizarre prediction using the laws of the quantum theory. He claimed that two uncharged parallel metal plates will attract each other, as if by magic. Normally parallel plates are stationary, since they lack any net charge. But the vacuum between the two parallel plates is not empty, but full of "virtual particles," which dart in and out of existence.

For brief periods of time, electron-antielectron pairs burst out of nothing, only to be annihilated and disappear back into the vacuum. Ironically, empty space, which was once thought to be devoid of anything, now turns out to be churning with quantum activity. Normally tiny bursts of matter and antimatter would seem to violate the conservation of energy. But because of the uncertainty principle, these tiny violations are incredibly short-lived, and on average energy is still conserved.

Casimir found that the cloud of virtual particles will create a net pressure in the vacuum. The space between the two parallel plates is confined, and hence the pressure is low. But the pressure outside the plates is unconfined and larger, and hence there will be a net pressure pushing the plates together.

Normally the state of zero energy occurs when these two plates are at rest and sitting far apart from each other. But as the plates come closer together, you can extract energy out of them. Thus, because kinetic energy has been taken out of the plates, the energy of the plates is less than zero.

This negative energy was actually measured in the laboratory in 1948, and the results confirmed Casimir's prediction. Thus, negative energy and the Casimir effect are no longer science fiction but established fact. The problem, however, is that the Casimir effect is quite small; it takes delicate, state-of-the-art measuring equipment to detect this energy in the laboratory. (In general, the Casimir energy is proportional to the inverse fourth power of the distance of separation between the plates. This means that the smaller the distance of separation, the larger the energy.) The Casimir effect was measured precisely in 1996 by Steven Lamoreaux at the Los Alamos National Laboratory, and the attractive force is 1/30,000 the weight of an ant.

Since Alcubierre first proposed his theory, physicists have discovered a number of strange properties. The people inside the starship are causally disconnected from the outside world. This means that you cannot simply press a button at will and travel faster than light. You cannot communicate through the bubble. There has to be a preexisting "highway" through space and time, like a series of trains passing by on a regular timetable. In this sense, the starship would not be an ordinary ship that can change directions and speeds at will. The starship would actually be like a passenger car riding on a preexisting "wave" of compressed space, coasting along a preexisting corridor of warped space-time. Alcubierre speculates, "We would need a series of generators of exotic matter along the way, like a highway, that manipulate space for you in a synchronized way."

Actually, even more bizarre types of solutions to Einstein's equations can be found. Einstein's equations state that if you are given a certain amount of mass or energy, you can compute the warping of space-time that the mass or energy will generate (in the same way that if you throw a rock into a pond, you can calculate the ripples that it will create). But you can also run the equations backward. You can start with a bizarre space-time, the kind found in episodes of *The Twilight Zone.* (In these universes, for example, you can open up a door and find yourself on the moon. You can run around a tree and find yourself backward in time, with your heart on the right side of your body.) Then you calculate the distribution of matter and energy associated with that particular space-time. (This means that if you are given a bizarre collection of waves on the surface of a pond, you can work backward and calculate the distribution of rocks necessary to produce these waves). This was, in fact, the way in which Alcubierre derived his equations. He began with a space-time consistent with going faster than light, and then he worked backward and calculated the energy necessary to produce it.

Wormholes and Black Holes

Besides stretching space, the second possible way to break the light barrier is by ripping space, via wormholes, passageways that connect two universes. In fiction, the first mention of a wormhole came from Oxford mathematician Charles Dodgson, who wrote *Through the Looking Glass* under the pen name Lewis Carroll. The Looking Glass of Alice is the wormhole, connecting the countryside of Oxford with the magical world of Wonderland. By placing her hand through the Looking Glass, Alice can be transported instantly from one universe to the next. Mathematicians call these "multiply connected spaces."

The concept of wormholes in physics dates back to 1916, one year after Einstein published his epic general theory of relativity. Physicist Karl Schwarzschild, then serving in the Kaiser's army, was able to solve Einstein's equations exactly for the case of a single pointlike star. Far from the star, its gravitational field was very similar to that of an ordinary star, and in fact Einstein used Schwarzschild's solution to calculate the deflection of light around a star. Schwarzschild's solution had an immediate and profound impact on astronomy, and even today it is one of the best-known solutions of Einstein's equations. For generations, physicists used the gravitational field around this pointlike star as an approximation to the field around a real star, which has a finite diameter.

But if you took this pointlike solution seriously, then lurking at the center of it was a monstrous pointlike object that has shocked and amazed physicists for almost a century–a black hole. Schwarzschild's solution for the gravity of a pointlike star was like a Trojan Horse. On the outside it looked like a gift from heaven, but on the inside there lurked all sorts of demons and ghosts. But if you accepted one, you had to accept the other. Schwarzschild's solution showed that as you approached this pointlike star, bizarre things happened. Surrounding the star was an invisible sphere (called the "event horizon") that was a point of no return. Everything checked in, but nothing could check out, like a Roach Motel. Once you passed through the event horizon, you never came back. (Once inside the event horizon, you would have to travel faster than light to escape back outside the event horizon, and that would be impossible.)

As you approached the event horizon, your atoms would be stretched by tidal forces. The gravity felt by your feet would be much greater than the gravity felt by your head, so you would be "spaghettified" and then ripped apart. Similarly, the atoms of your body would also be stretched and torn apart by gravity.

To an outside observer watching you approach the event horizon, it would appear that you were slowing down in time. In fact, as you hit the event horizon, it would appear that time had stopped!

Furthermore, as you fell past the event horizon, you would see light that has been trapped and circulating around this black hole for billions of years. It would seem as if you were watching a motion picture film, detailing the entire history of the black hole, going back to its very origin.

And finally, if you could fall straight through to the black hole, there would be another universe on the other side. This is called the Einstein-Rosen Bridge, first introduced by Einstein in 1935; it is now called a wormhole.

Einstein and other physicists believed a star could never evolve naturally into such a monstrous object. In fact, in 1939 Einstein published a paper showing that a circulating mass of gas and dust will never condense into such a black hole. So although there was a wormhole lurking in the center of a black hole, he was confident that such a strange object could never form by natural means. In fact, astrophysicist Arthur Eddington once said that there should "be a law of nature to prevent a star from behaving in this absurd way." In other words, the black hole was indeed a legitimate solution of Einstein's equations, but there was no known mechanism that could form one by natural means.

All this changed with the advent of a paper by J. Robert Oppenheimer and his student Hartland Snyder, written that same year, showing that black holes can indeed be formed by natural means. They assumed that a dying star had used up its nuclear fuel and then collapsed under gravity, so that it imploded under its own weight. If gravity could compress the star to within its event horizon, then nothing known to science could prevent gravity from squeezing the star to a point-particle, the black hole. (This implosion method may have given Oppenheimer the clue for building the Nagasaki bomb just a few years later, which depends on imploding a sphere of plutonium.)

The next breakthrough came in 1963, when New Zealand mathematician Roy Kerr examined perhaps the most realistic example of a black hole. Objects spin faster as they shrink, in much the same way that skaters spin faster when they bring in their arms close to their body. As a result black holes should be spinning at fantastic rates.

Kerr found that a spinning black hole would not collapse into a pointlike star, as Schwarzschild assumed, but would collapse into a spinning ring. Anyone unfortunate enough to hit the ring would perish; but someone falling into the ring would not die, but would actually fall through. But instead of winding up on the other side of the ring, he or she would pass through the Einstein-Rosen Bridge and wind up in another universe. In other words, the spinning black hole is the rim of Alice's Looking Glass.

If he or she were to move around the spinning ring a second time, he or she would enter yet another universe. In fact, repeated entry into the spinning ring would put a person in different parallel universes, much like hitting the "up" button on an elevator. In principle, there could be an infinite number of universes, each stacked on top of each other. "Pass through this magic ring and–presto!–you're in a completely different universe where radius and mass are negative!" Kerr wrote.

There is an important catch, however. Black holes are examples of "nontransversable wormholes"; that is, passing through the event horizon is a one-way trip. Once you pass through the event horizon and the Kerr ring, you cannot go backward through the ring and out through the event horizon.

But in 1988 Kip Thorne and colleagues at Cal Tech found an example of a transversable wormhole, that is, one through which you could pass freely back and forth. In fact, for one solution, the travel through a wormhole would be no worse than riding on an airplane.

Normally gravity would crush the throat of the wormhole, destroying the astronauts trying to reach the other side. That is one reason that faster-than-light travel through a wormhole is not possible. But the repulsive force of negative energy or negative mass could conceivably keep the throat open sufficiently long to allow astronauts a clear passage. In other words, negative mass or energy is essential for both the Alcubierre drive and the wormhole solution.

In the last few years an astonishing number of exact solutions have been found to Einstein's equations that allow for wormholes. But do wormholes really exist, or are they just a figment of mathematics? There are several major problems facing wormholes.

First, to create the violent distortions of space and time necessary to travel through a wormhole, one would need fabulous amounts of positive and negative matter, on the order of a huge star or a black hole. Matthew Visser, a physicist at Washington University, estimates that the amount of negative energy you would need to open up a 1-meter wormhole is comparable to the mass of Jupiter, except that it would need to be negative. He says, "You need about minus one Jupiter mass to do the job. Just manipulating a positive Jupiter mass of energy is already pretty freaky, well beyond our capabilities into the foreseeable future."

Kip Thorne of the California Institute of Technology speculates that "it will turn out that the laws of physics do allow sufficient exotic matter in wormholes of human size to hold the wormhole open. But it will also turn out that the technology for making wormholes and holding them open is unimaginably far beyond the capabilities of our human civilization."

Second, we do not know how stable these wormholes would be. The radiation generated by these wormholes might kill anyone who enters. Or perhaps the wormholes would not be stable at all, closing as soon as one entered them.

Third, light beams falling into the black hole would be blue shifted; that is, they would attain greater and greater energy as they came close to the event horizon. In fact, at the event horizon itself, light is technically infinitely blue shifted, so the radiation from this infalling energy could kill anyone in a rocket.

Let us discuss these problems in some detail. One problem is to amass enough energy to rip the fabric of space and time. The simplest way to do this is to compress an object until it becomes smaller than its "event horizon." For the sun, this means compressing it down to about 2 miles in diameter, whereupon it will collapse into a black hole. (The Sun's gravity is too weak to compress it naturally down to 2 miles, so our sun will never become a black hole. In principle, this means that anything, even you, can become a black hole if you were sufficiently compressed. This would mean compressing all the atoms of your body to smaller than subatomic distances–a feat that is beyond the capabilities of modern science.)

A more practical approach would be to assemble a battery of laser beams to fire an intense beam at a specific spot. Or to build a huge atom smasher to create two beams, which would then collide with each other at fantastic energies, sufficient to create a small tear in the fabric of space-time.

Planck Energy and Particle Accelerators

One can calculate the energy necessary to create an instability in space and time: it is of the order of the Planck energy, or 10^{19} billion electron volts. This is truly an unimaginably large number, a quadrillion times larger than the energy attainable with today's most powerful machine, the Large Hadron Collider (LHC), located outside Geneva, Switzerland. The LHC is capable of swinging protons in a large "doughnut" until they reach energies of trillions of electron volts, energies not seen since the big bang. But even this monster of a machine falls far short of producing energy anywhere near the Planck energy.

The next particle accelerator after the LHC will be the International Linear Collider (ILC). Instead of bending the path of subatomic particles into a circle, the ILC will shoot them down a straight path. Energy will be injected as the particles move along this path, until they attain unimaginably large energies. Then a beam of electrons will collide with antielectrons, creating a huge burst of energy. The ILC will be 30 to 40 kilometers long, or ten times the length of the Stanford Linear Accelerator, currently the largest linear accelerator. If all goes well, the ILC is due to be completed sometime in the next decade.

The energy produced by the ILC will be .5 to 1.0 trillion electron volts–less than the 14 trillion electron volts of the LHC, but this is deceptive. (In the LHC, the collisions between the protons take place between the constituent quarks making up the proton. Hence the collisions involving the quarks are less than 14 trillion electron volts. That is why the ILC will produce collision energies larger than those of the LHC.) Also, because the electron has no known constituent, the dynamics of the collisions between electron and antielectron are simpler and cleaner.

But realistically, the ILC, too, falls far short of being able to open up a hole in space-time. For that, you would need an accelerator a quadrillion times more powerful. For our Type 0 civilization, which uses dead plants for fuel (e.g., oil and coal), this technology is far beyond anything we can muster. But it may become possible for a Type III civilization.

Remember, a Type III civilization, which is galactic in its use of energy, consumes 10 billion times more energy than a Type II civilization, whose consumption is based on the energy of a single star. And a Type II civilization in turn consumes 10 billion times more energy than a Type I civilization, whose consumption is based on the energy of a single planet. In one hundred to two hundred years, our feeble Type 0 civilization will reach Type I status.

Given that projection, we are a long, long way from being able to achieve the Planck energy. Many physicists believe that at extremely tiny distances, at the Planck distance of 10^{-33} centimeters, space is not empty or smooth but becomes "foamy"; it is frothing with tiny bubbles that constantly pop into existence, collide with other bubbles, and then vanish back into the vacuum. These bubbles that dart in and out of the vacuum are "virtual universes," very similar to the virtual particles of electrons and antielectrons that pop into existence and then disappear.

Normally, this quantum space-time "foam" is completely invisible to us. These bubbles form at such tiny distances that we cannot observe them. But quantum physics suggests that if we concentrate enough energy at a single point, until we reach the Planck energy, these bubbles can become large. Then we would see space-time frothing with tiny bubbles, each bubble a wormhole connected to a "baby universe."

In the past these baby universes were considered an intellectual curiosity, a strange consequence of pure mathematics. But now physicists are seriously thinking that our universe might have originally started off as one of these baby universes.

Such thinking is sheer speculation, but the laws of physics allow for the possibility of opening a hole in space by concentrating enough energy at a single point, until we access the space-time foam and wormholes emerge connecting our universe to a baby universe.

Achieving a hole in space would, of course, require major breakthroughs in our technology, but again, it might be possible for a Type III civilization. For example, there have been promising developments in something called a "Wakefield tabletop accelerator." Remarkably, this atom smasher is so small that it can be placed on top of a table yet generate billions of electron volts of energy. The Wakefield tabletop accelerator works by firing lasers onto charged particles, which then ride on the energy of that laser. Experiments done at the Stanford Linear Accelerator Center, the Rutherford Appleton Laboratory in England, and the École Polytechnique in Paris show that enormous accelerations are possible over small distances using laser beams and plasma to inject energy.

Yet another breakthrough was made in 2007, when physicists and engineers at the Stanford Linear Accelerator Center, UCLA, and USC demonstrated that you can double the energy of a huge particle accelerator in just 1 meter. They started with a beam of electrons that are fired down a 2-mile-long tube in Stanford, reaching an energy of 42 billion electron volts. Then these high-energy electrons were sent through an "afterburner," which consisted of a plasma chamber only 88 centimeters long, where the electrons pick up an additional 42 billion electron volts, doubling their energy. (The plasma chamber is filled with lithium gas. As the electrons pass through the gas, they create a plasma wave that creates a wake. This wake in turn flows to the back of the electron beam and then shoves it forward, giving it an extra boost.) In this stunning achievement, the physicists improved by a factor of three thousand the previous record for the amount of energy per meter they could accelerate an electron beam. By adding such "afterburners" to existing accelerators, one might in principle double their energy, almost for free.

Today the world record for a Wakefield tabletop accelerator is 200 billion electron volts per meter. There are numerous problems scaling this result to longer distances (such as maintaining the stability of the beam as laser power is pumped into it). But assuming that we could maintain a power level of 200 billion electron volts per meter, this means that an accelerator capable of reaching the Planck energy would have to be 10 light-years long. This is well within the capability of a Type III civilization.

Wormholes and stretched space may give us the most realistic way of breaking the light barrier. But it is not known if these technologies are stable; if they are, it would still take a fabulous amount of energy, positive or negative, to make them work.

Perhaps an advanced Type III civilization might already have this technology. It might be millennia before we can even think about harnessing power on this scale. Because there is still controversy over the fundamental laws governing the fabric of space-time at the quantum level, I would classify this as a Class II impossibility.

12 : TIME TRAVEL

If time travel is possible, then where are the
tourists from the future?

—STEPHEN HAWKING

"[Time travel] is against reason," said Filby.
"What reason?" said the Time Traveler.

—H. G. WELLS

In the novel *Janus Equation,* writer G. Spruill explored one of the harrowing problems with time travel. In this tale a brilliant mathematician whose goal is to discover the secret of time travel meets a strange, beautiful woman, and they become lovers, although he knows nothing about her past. He becomes intrigued about finding out her true identity. Eventually he discovers that she once had plastic surgery to change her features. And that she had a sex change operation. Finally, he discovers that "she" is actually a time traveler from the future, and that "she" is actually himself, but from the future. This means that he made love to himself. And one is left wondering, what would have happened if they had had a child? And if this child went back into the past, to grow up to become the mathematician at the beginning of the story, then is it possible to be your own mother and father and son and daughter?

CHANGING THE PAST

Time is one of the great mysteries of the universe. We are all swept up in the river of time against our will. Around AD 400, Saint Augustine wrote extensively about the paradoxical nature of time: "How can the past and future be, when the past no longer is, and the future is not yet? As for the present, if it were always present and never moved on to become the past, it would not be time, but eternity." If we take Saint Augustine's logic further, we see that time is not possible, since the past is gone, the future does not exist, and the present exists only for an instant. (Saint Augustine then asked profound theological questions about how time must influence God, questions that are relevant even today. If God is omnipotent and all-powerful, he wrote, then is He bound by the passing of time? In other words, does God, like the rest of us mortals, have to rush because He is late for an appointment? Saint Augustine eventually concluded that God is omnipotent and hence cannot be constrained by time and would therefore have to exist "outside of time." Although the concept of being outside of time seems absurd, it's one idea that is recurring in modern physics, as we will see.)

Like Saint Augustine, all of us have at some time wondered about the strange nature of time and how it differs from space. If we can move forward and backward in space, why not in time? All of us have also wondered what the future may hold for us, in the time beyond our years. Humans have a finite lifetime, but we are intensely curious about events that will happen long after we are gone.

Although our longing to travel in time is probably as ancient as humanity, apparently the very first written time travel story is *Memoirs of the Twentieth Century*, written in 1733 by Samuel Madden, about an angel from the year 1997 who journeys over 250 years into the past to give documents to a British ambassador that describe the world of the future.

There would be many more such stories. The 1838 short story "Missing One's Coach: An Anachronism," written anonymously, is about a person waiting for a coach who suddenly finds himself a thousand years in the past. He meets a monk from an ancient monastery and tries to explain to him how history will progress for the next thousand years. Afterward he suddenly finds himself just as mysteriously transported back to the present, except that he has missed his coach.

Even the 1843 Charles Dickens novel, *A Christmas Carol,* is a kind of time travel story, since Ebenezer Scrooge is taken into the past and into the future to witness the world before the present and after his death.

In American literature the first appearance of time travel dates back to Mark Twain's 1889 novel, *A Connecticut Yankee in King Arthur's Court.* A nineteenth-century Yankee is wrenched backward through time to wind up in King Arthur's court in AD 528. He is taken prisoner and is about to be burned at the stake, but then he declares he has the power to blot out the sun, knowing that an eclipse of the sun would happen on that very day. When the sun is eclipsed, the mob is horrified and agrees to set him free and grant him privileges in exchange for the return of the sun.

But the first serious attempt to explore time travel in fiction was H. G. Wells's classic *The Time Machine,* in which the hero is sent hundreds of thousands of years into the future. In that distant future, humanity itself has genetically split into two races, the menacing Moorlocks who maintain the grimy underground machines, and the useless, childlike Eloi who dance in the sunlight in the world above, never realizing their awful fate (to be eaten by the Moorlocks).

Since then, time travel has become a regular feature of science fiction, from *Star Trek* to *Back to the Future.* In *Superman I,* when Superman learns that Lois Lane has died, he decides in desperation to turn back the hands of time, rocketing himself around the Earth, faster than the speed of light, until time itself goes backward. The Earth slows down, stops, and eventually spins in the opposite direction, until all clocks on the Earth beat backward. Floodwaters rage backward, broken dams miraculously heal themselves, and Lois Lane comes back from the dead.

From the perspective of science, time travel was impossible in Newton's universe, where time was seen as an arrow. Once fired, it could never deviate from its past. One second on the Earth was one second throughout the universe. This conception was overthrown by Einstein, who showed that time was more like a river that meandered across the universe, speeding up and slowing down as it snaked across stars and galaxies. So one second on the Earth is not absolute; time varies when we move around the universe.

As I discussed earlier, according to Einstein's special theory of relativity, time slows down inside a rocket the faster it moves. Science fiction writers have speculated that if you could break the light barrier, you could go back in time. But this is not possible, since you would have to have infinite mass in order to reach the speed of light. The speed of light is the ultimate barrier for any rocket. The crew of the *Enterprise* in *Star Trek IV: The Voyage Home* hijacked a Klingon spaceship and used it to whip around the sun like a slingshot and break the light barrier to wind up in San Francisco in the 1960s. But this defies the laws of physics.

Nonetheless, time travel to the future is possible, and has been experimentally verified millions of times. The journey of the hero of *The Time Machine* into the far future is actually physically possible. If an astronaut were to travel near the speed of light, it might take him, say, one minute to reach the nearest stars. Four years would have elapsed on the Earth, but for him only one minute would have passed, because time would have slowed down inside the rocket ship. Hence he would have traveled four years into the future, as experienced here on Earth. (Our astronauts actually take a short trip into the future every time they go into outer space. As they travel at 18,000 miles per hour above the Earth, their clocks beat a tiny bit slower than clocks on the Earth. Hence, after a yearlong mission on the space station, they have actually journeyed a fraction of a second into the future by the time they land back on Earth. The world record for traveling into the future is currently held by Russian cosmonaut Sergei Avdeyev, who orbited for 748 days and was hence hurled .02 seconds into the future.)

So a time machine that can take us into the future is consistent with Einstein's special theory of relativity. But what about going backward in time?

If we could journey back into the past, history would be impossible to write. As soon as a historian recorded the history of the past, someone could go back into the past and rewrite it. Not only would time machines put historians out of business, but they would enable us to alter the course of time at will. If, for example, we were to go back to the era of the dinosaurs and accidentally step on a mammal that happens to be our ancestor, perhaps we would accidentally wipe out the entire human race. History would become an unending, madcap Monty Python episode, as tourists from the future trampled over historic events while trying to get the best camera angle.

Time Travel: Physicists' Playground

Perhaps the person who has distinguished himself the most on the dense mathematical equations of black holes and time machines is cosmologist Stephen Hawking. Unlike other students of relativity who often distinguish themselves in mathematical physics at an early age, Hawking was actually not an outstanding student as a youth. He was obviously extremely bright, but his teachers would often notice that he was not focused on his studies and never lived up to his full potential. But a turning point came in 1962, after he graduated from Oxford, when he first began to notice the symptoms of ALS (amyotrophic lateral sclerosis, or Lou Gehrig's disease). He was rocked by the news that he was suffering from this incurable motor neuron disease that would rob him of all motor functions and likely soon kill him. At first the news was extremely upsetting. What would be the use of getting a Ph.D. if he was going to die soon anyway?

But once he got over the initial shock he became focused for the first time in his life. Realizing that he did not have long to live, he began to ferociously tackle some of the most difficult problems in general relativity. In the early 1970s he published a landmark series of papers showing that "singularities" in Einstein's theory (where the gravitational field becomes infinite, like at the center of black holes and at the instant of the big bang) were an essential feature of relativity and could not be easily dismissed (as Einstein thought). In 1974 Hawking also proved that black holes are not entirely black, but gradually emit radiation, now known as Hawking radiation, because radiation can tunnel through the gravity field of even a black hole. This paper was the first major application of the quantum theory to relativity theory, and it represents his best known work.

As predicted, ALS slowly led to paralysis of his hands, legs, and even his vocal cords, but at a much slower rate than the doctors had originally predicted. As a result, he has passed many of the usual milestones of normal people, fathering three children (he is now a grandfather), divorcing his first wife in 1991, four years later marrying the wife of the man who created his voice synthesizer, and filing for divorce from his second wife in 2006. In 2007 he made headlines when he went aboard a jet airplane that sent him into weightlessness, fulfilling a lifelong wish of his. His next goal is to blast off into outer space.

Today he is almost totally paralyzed in his wheelchair, communicating to the outside world via movements of his eyes. Yet even with this crushing disability, he still cracks jokes, writes papers, gives lectures, and engages in controversy. He is more productive moving his two eyes than are teams of scientists who have full control over their bodies. (His colleague at Cambridge University, Sir Martin Rees, who was appointed Astronomer Royal by the Queen, once confided to me that Hawking's disability does prevent him from doing the tedious calculations necessary to keep at the top of his game. So instead he concentrates on generating new and fresh ideas rather than cranking out difficult calculations, which can be done by his students.)

In 1990 Hawking read papers of his colleagues proposing their version of a time machine, and he was immediately skeptical. His intuition told him that time travel was not possible because there are no tourists from the future. If time travel were as common as taking a Sunday picnic in the park, then time travelers from the future should be pestering us with their cameras, asking us to pose for their picture albums.

Hawking also raised a challenge to the world of physics. There ought to be a law, he proclaimed, making time travel impossible. He proposed a "Chronology Protection Conjecture" to ban time travel from the laws of physics in order to "make history safe for historians."

The embarrassing thing, however, was that no matter how hard physicists tried, they could not find a law to prevent time travel. Apparently time travel seems to be consistent with the known laws of physics. Unable to find any physical law that makes time travel impossible, Hawking recently changed his mind. He made headlines in the London papers when he said, "Time travel may be possible, but it is not practical."

Once considered to be fringe science, time travel has suddenly become a playground for theoretical physicists. Physicist Kip Thorne of Cal Tech writes, "Time travel was once solely the province of science fiction writers. Serious scientists avoided it like the plague–even when writing fiction under pseudonyms or reading it in privacy. How times have changed! One now finds scholarly analyses of time travel in serious scientific journals, written by eminent theoretical physicists... Why the change? Because we physicists have realized that the nature of time is too important an issue to be left solely in the hands of science fiction writers."

The reason for all this confusion and excitement is that Einstein's equations allow for many kinds of time machines. (Whether they will survive the challenges from the quantum theory, however, is still in doubt.) In Einstein's theory, in fact, we often encounter something called "closed time-like curves," which is the technical term for paths that allow for time travel into the past. If we followed the path of a closed time-like curve, we would set out on a journey and return before we left.

The first time machine involves a wormhole. There are many solutions of Einstein's equations that connect two distant points in space. But since space and time are intimately intertwined in Einstein's theory, this same wormhole can also connect two points in time. By falling down the wormhole, you could journey (at least mathematically) into the past. Conceivably, you could then journey to the original starting point and meet yourself before you left. But as we mentioned in the previous chapter, passing through the wormhole at the center of a black hole is a one-way trip. As physicist Richard Gott has said, "I don't think there's any question that a person could travel back in time while in a black hole. The question is whether he could ever emerge to brag about it."

Another time machine involves a spinning universe. In 1949 mathematician Kurt Gödel found the first solution of Einstein's equations involving time travel. If the universe spins, then, if you traveled around the universe fast enough, you might find yourself in the past and arrive before you left. A trip around the universe is therefore also a trip into the past. When astronomers would visit the Institute for Advanced Study, Gödel would often ask them if they ever found evidence that the universe was spinning. He was disappointed when they told him that there was clearly evidence that the universe expanded, but the net spin of the universe was probably zero. (Otherwise, time travel might be commonplace, and history as we know it would collapse.)

Third, if you walk around an infinitely long, rotating cylinder, you also might arrive before you left. (This solution was found by W. J. van Stockum in 1936, before Gödel's time traveling solution, but van Stockum was apparently unaware that his solution allowed for time travel.) In this case, if you danced around a spinning May Pole on May Day, you might find yourself in the month of April. (The problem with this design, however, is that the cylinder must be infinite in length and spin so fast that most materials would fly apart.)

The most recent example of time travel was found by Richard Gott of Princeton in 1991. His solution was based on finding gigantic cosmic strings (which may be leftovers from the original big bang). He assumed that two large cosmic strings were about to collide. If you quickly traveled around these colliding cosmic strings, you would travel back in time. The advantage of this type of time machine is that you would not need infinite spinning cylinders, spinning universes, or black holes. (The problem with this design, however, is that you must first find huge cosmic strings floating in space and then make them collide in a precise fashion. And the possibility of going back in time would last only a brief period.) Gott says, "A collapsing loop of string large enough to allow you to circle it once and go back in time a year would have to have more than half the mass-energy of an entire galaxy."

But the most promising design for a time machine is the "transversable wormhole," mentioned in the last chapter, a hole in space-time in which a person could freely walk back and forth in time. On paper, transversable wormholes can provide not only faster-than-light travel, but also travel in time. The key to transversable wormholes is negative energy.

A transversable wormhole time machine would consist of two chambers. Each chamber would consist of two concentric spheres, which would be separated by a tiny distance. By imploding the outer sphere, the two spheres would create a Casimir effect and hence negative energy. Assume that a Type III civilization is able to string a wormhole between these two chambers (possibly extracting one from the space-time foam). Next, take the first chamber and send it into space at near light-speed velocities. Time slows down in that chamber, so the two clocks are no longer in synchronization. Time beats at different rates inside the two chambers, which are connected by a wormhole.

If you are in the second chamber, you can instantly pass through the wormhole to the first chamber, which exists at an earlier time. Thus you have gone backward in time.

There are formidable problems facing this design. The wormhole may be quite tiny, much smaller than an atom. And the plates may have to be squeezed down to Planck-length distances to create enough negative energy. Lastly, you would be able to go back in time only to the point when the time machines were built. Before then, time in the two chambers would be beating at the same rate.

Paradoxes and Time Conundrums

Time travel poses all sorts of problems, both technical as well as social. The moral, legal, and ethical issues are raised by Larry Dwyer, who notes, "Should a time traveler who punches his younger self (or vice versa) be charged with assault? Should the time traveler who murders someone and then flees into the past for sanctuary be tried in the past for crimes he committed in the future? If he marries in the past can he be tried for bigamy even though his other wife will not be born for almost 5,000 years?"

But perhaps the thorniest problems are the logical paradoxes raised by time travel. For example, what happens if we kill our parents before we are born? This is a logical impossibility. It is sometimes called the "grandfather paradox."

There are three ways to resolve these paradoxes. First, perhaps you simply repeat past history when you go back in time, therefore fulfilling the past. In this case, you have no free will. You are forced to complete the past as it was written. Thus, if you go back into the past to give the secret of time travel to your younger self, then it was meant to happen that way. The secret of time travel came from the future. It was destiny. (But this does not tell us where the original idea came from.)

Second, you have free will, so you can change the past, but within limits. Your free will is not allowed to create a time paradox. Whenever you try to kill your parents before you are born, a mysterious force prevents you from pulling the trigger. This position has been advocated by the Russian physicist Igor Novikov. (He argues that there is a law preventing us from walking on the ceiling, although we might want to. Hence there might be a law preventing us from killing our parents before we are born. Some strange law prevents us from pulling the trigger.)

Third, the universe splits into two universes. On one time line the people whom you killed look just like your parents, but they are different, because you are now in a parallel universe. This latter possibility seems to be the one consistent with the quantum theory, as I will discuss later when I talk about the multiverse.

The second possibility is explored in the movie *Terminator 3,* in which Arnold Schwarzenegger plays a robot from the future where murderous machines have taken over. The few remaining humans, hunted down like animals by the machines, are led by a great leader whom the machines have been unable to kill. Frustrated, the machines send a series of killer robots back to the past, before the great leader was born, to kill off his mother. But after epic battles, human civilization is eventually destroyed at the end of the movie, as it was meant to be.

Back to the Future explored the third possibility. Dr. Brown invents a plutonium-fired DeLorean car, actually a time machine for traveling to the past. Michael J. Fox (Marty McFly) enters the machine and goes back and meets his teenage mother, who then falls in love with him. This poses a sticky problem. If Marty McFly's teenage mother spurns his future father, then they never would have married, and Michael J. Fox's character would never have been born.

The problem is clarified a bit by Doc Brown. He goes to the blackboard and draws a horizontal line, representing the time line of our universe. Then he draws a second line, which branches off the first line, representing a parallel universe that opens up when you change the past. Thus, whenever we go back into the river of time, the river forks into two rivers, and one time line becomes two time lines, or what is called the "many worlds" approach, which we will discuss in the next chapter.

This means that all time travel paradoxes can be solved. If you have killed your parents before you were born, it simply means you have killed some people who are genetically identical to your parents, with the same memories and personalities, but they are not your true parents.

The "many worlds" idea solves at least one main problem with time travel. To a physicist, the number one criticism of time travel (besides finding negative energy) is that radiation effects will build up until either you are killed the instant you enter the machine or the wormhole collapses on you. Radiation effects build up because any radiation entering the time portal will be sent back into the past, where it will eventually wander around the universe until it reaches the present day, and then it will fall into the wormhole again. Since radiation can enter the mouth of the wormhole an infinite number of times, the radiation inside the wormhole can become incredibly strong–strong enough to kill you. But the "many worlds" interpretation solves this problem. If the radiation goes into the time machine and is sent into the past, it then enters a new universe; it cannot reenter the time machine again, and again, and again. This simply means that there are an infinite number of universes, one for each cycle, and each cycle contains just one photon of radiation, not an infinite amount of radiation.

In 1997, the debate was clarified a bit when three physicists finally proved that Hawking's program to ban time travel was inherently flawed. Bernard Kay, Marek Radzikowski, and Robert Wald showed that time travel was consistent with all the known laws of physics, except in one place. When traveling in time, all the potential problems were concentrated at the event horizon (located near the entrance to the wormhole). But the horizon is precisely where we expect Einstein's theory to break down and quantum effects to take over. The problem is that whenever we try to calculate radiation effects as we enter a time machine, we have to use a theory that combines Einstein's theory of general relativity with the quantum theory of radiation. But whenever we naïvely try to marry these two theories, the resulting theory makes no sense: it yields a series of infinite answers that are meaningless.

This is where a theory of everything takes over. All problems of traveling through a wormhole that have bedeviled physicists (e.g., the stability of the wormhole, the radiation that might kill you, the closing of the wormhole as you entered it) are concentrated at the event horizon, precisely where Einstein's theory made no sense.

Thus the key to understanding time travel is to understand the physics of the event horizon, and only a theory of everything can explain this. This is the reason that most physicists today would agree that one way to definitively settle the time travel question is to come up with a complete theory of gravity and space-time.

A theory of everything would unite the four forces of the universe and enable us to calculate what would happen when we entered a time machine. Only a theory of everything could successfully calculate all the radiation effects created by a wormhole and definitively settle the question of how stable wormholes would be when we entered the time machine. And even then, we might have to wait for centuries or even longer to actually build a machine to test these theories.

Because the laws of time travel are so closely linked to the physics of wormholes, time travel seems to qualify as a Class II impossibility.

13: PARALLEL UNIVERSES

"But do you really mean, sir," said Peter, "that there could be other worlds–all over the place, just around the corner–like that?"

"Nothing is more probable," said the Professor . . . while he muttered to himself, "I wonder what they do teach them at these schools."

–C. S. LEWIS, *THE LION, THE WITCH AND THE WARDROBE*

listen: there's a hell of a good universe next door; let's go

–E. E. CUMMINGS

Are alternate universes really possible? They are a favorite device for Hollywood scriptwriters, as in the *Star Trek* episode called "Mirror, Mirror." Captain Kirk is accidentally transported to a bizarre parallel universe in which the Federation of Planets is an evil empire held together by brutal conquest, greed, and plunder. In that universe Spock wears a menacing beard and Captain Kirk is the leader of a band of ravenous pirates, advancing by enslaving their rivals and assassinating their superiors.

Alternate universes enable us to explore the world of "what if" and its delicious, intriguing possibilities. In the *Superman* comics, for example, there have been several alternate universes in which Superman's home planet, Krypton, never blew up, or Superman finally reveals his true identity as mild-mannered Clark Kent, or he marries Lois Lane and has superkids. But are parallel universes just the domain of *Twilight Zone* reruns, or do they have a basis in modern physics?

Throughout history going back to almost all ancient societies, people have believed in other planes of existence, the homes of the gods or ghosts. The Church believes in heaven, hell, and purgatory. The Buddhists have Nirvana and different states of consciousness. And the Hindus have thousands of planes of existence.

Christian theologians, at a loss to explain where heaven might be located, have often speculated that perhaps God lives in a higher dimensional plane. Surprisingly, if higher dimensions did exist, many of the properties ascribed to the gods might become possible. A being in a higher dimension might be able to disappear and reappear at will or walk through walls–powers usually ascribed to deities.

Recently the idea of parallel universes has become one of the most hotly debated topics in theoretical physics. There are, in fact, several types of parallel universes that force us to reconsider what we mean by what is "real." What is at stake in this debate about various parallel universes is nothing less than the meaning of reality itself.

There are at least three types of parallel universes that are intensely discussed in the scientific literature:

a. hyperspace, or higher dimensions,
b. the multiverse, and
c. quantum parallel universes.

Hyperspace

The parallel universe that has been the subject of the longest historical debate is one of higher dimensions. The fact that we live in three dimensions (length, width, height) is common sense. No matter how we move an object in space, all positions can be described by these three coordinates. In fact, with these three numbers we can locate any object in the universe, from the tip of our noses to the most distant of all galaxies.

A fourth spatial dimension seems to violate common sense. If smoke, for example, is allowed to fill up a room, we do not see the smoke disappearing into another dimension. Nowhere in our universe do we see objects suddenly disappearing or drifting off into another universe. This means that any higher dimensions, if they exist at all, must be smaller than an atom.

Three spatial dimensions form the fundamental basis of Greek geometry. Aristotle, for example, in his essay "On Heaven," wrote, "The line has magnitude in one way, the plane in two ways, and the solid in three ways, and beyond these there is no other magnitude because the three are all." In AD 150 Ptolemy of Alexandria offered first "proof" that higher dimensions were "impossible." In his essay "On Distance," he reasoned as follows. Draw three lines that are mutually perpendicular (like the lines forming the corner of a room). Clearly, he said, a fourth line perpendicular to the other three cannot be drawn, hence a fourth dimension must be impossible. (What he actually proved was that our brains are incapable of visualizing the fourth dimension. The PC on your desk calculates in hyperspace all the time.)

For two thousand years, any mathematician who dared to speak of the fourth dimension potentially suffered ridicule. In 1685 mathematician John Wallis polemicized against the fourth dimension, calling it a "Monster in Nature, less possible than a Chimera or Centaure." In the nineteenth century Karl Gauss, the "prince of mathematicians," worked out much of the mathematics of the fourth dimension but was afraid to publish because of the backlash it would cause. But privately Gauss conducted experiments to test whether flat, three-dimensional Greek geometry really described the universe. In one experiment he placed his assistants on three mountaintops. Each one had a lantern, thereby forming a huge triangle. Gauss then measured the angles of each corner of the triangle. To his disappointment, he found that the interior angles all summed up to 180 degrees. He concluded that if there were deviations to standard Greek geometry, they must be so small that they could not be detected with his lanterns.

Gauss left it to his student, Georg Bernhard Riemann, to write down the fundamental mathematics of higher dimensions (which were then imported wholesale decades later into Einstein's theory of general relativity). In one powerful sweep, in a celebrated lecture Riemann delivered in 1854, he overthrew two thousand years of Greek geometry and established the basic mathematics of the higher, curved dimensions that we use even today.

After Riemann's remarkable discovery was popularized in Europe in the late 1800s, the "fourth dimension" became quite a sensation among artists, musicians, writers, philosophers, and painters. Picasso's cubism, in fact, was partly inspired by the fourth dimension, according to art historian Linda Dalrymple Henderson. (Picasso's paintings of women with eyes facing forward and nose to the side was an attempt to visualize a fourth-dimensional perspective, since one looking down from the fourth dimension could see a woman's face, nose, and the back of her head simultaneously.) Henderson writes, "Like a Black Hole, the 'fourth dimension' possessed mysterious qualities that could not be completely understood, even by the scientists themselves. Yet, the impact of 'the fourth dimension' was far more comprehensive than that of Black Holes or any other more recent scientific hypothesis except Relativity Theory after 1919."

Other painters drew from the fourth dimension, as well. In Salvador Dali's *Christus Hypercubius,* Christ is crucified in front of a strange, floating three-dimensional cross, which is actually a "tesseract," an unraveled four-dimensional cube. In his famous *Persistence of Memory,* he attempted to represent time as the fourth dimension, and hence the metaphor of melted clocks. Marcel Duchamps's *Nude Descending a Staircase* was an attempt to represent time as the fourth dimension by capturing the time-lapse motion of a nude walking down a staircase. The fourth dimension even pops up in a story by Oscar Wilde, "The Canterville Ghost," in which a ghost haunting a house lives in the fourth dimension.

The fourth dimension also appears in several of H. G. Wells's works, including *The Invisible Man, The Plattner Story,* and *The Wonderful Visit.* (In the latter, which has since been the basis of scores of

Hollywood movies and science fiction novels, our universe somehow collides with a parallel universe. A poor angel from the other universe falls into our universe after being accidentally shot by a hunter. Horrified by all the greed, pettiness, and selfishness of our universe, the angel eventually commits suicide.)

The idea of parallel universes was also explored, tongue-in-cheek, by Robert Heinlein in *The Number of the Beast.* Heinlein imagines a group of four brave individuals who romp across parallel universes in a mad professor's interdimensional sports car.

In the TV series *Sliders,* a young boy reads a book and gets the inspiration to build a machine that would allow him to "slide" between parallel universes. (The book that the young boy was reading was actually my book, *Hyperspace.*)

But historically the fourth dimension has been considered a mere curiosity by physicists. No evidence has ever been found for higher dimensions. This began to change in 1919 when physicist Theodor Kaluza wrote a highly controversial paper that hinted at the presence of higher dimensions. He started with Einstein's theory of general relativity, but placed it in five dimensions (one dimension of time and four dimensions of space; since time is the fourth space-time dimension, physicists now refer to the fourth spatial dimension as the fifth dimension). If the fifth dimension were made smaller and smaller, the equations magically split into two pieces. One piece describes Einstein's standard theory of relativity, but the other piece becomes Maxwell's theory of light!

This was a stunning revelation. Perhaps the secret of light lies in the fifth dimension! Einstein himself was shocked by this solution, which seemed to provide an elegant unification of light and gravity. (Einstein was so shaken by Kaluza's proposal that he mulled it over for two years before finally agreeing to have this paper published.) Einstein wrote to Kaluza, "The idea of achieving [a unified theory] by means of a five-dimensional cylinder world never dawned on me . . . At first glance, I like your idea enormously . . . The formal unity of your theory is startling."

For years physicists had asked the question: if light is a wave, then what is waving? Light can pass through billions of light-years of empty space, but empty space is a vacuum, devoid of any material. So what is waving in the vacuum? With Kaluza's theory we had a concrete proposal to answer this problem: light is ripples in the fifth dimension. Maxwell's equations, which accurately describe all the properties of light, emerge simply as the equations for waves traveling in the fifth dimension.

Imagine fish swimming in a shallow pond. They might never suspect the presence of a third dimension, because their eyes point to the side, and they can only swim forward and backward, left and right. A third dimension to them might appear impossible. But then imagine it rains on the pond. Although they cannot see the third dimension, they can clearly see the shadows of the ripples on the surface of the pond. In the same way, Kaluza's theory explained light as ripples traveling on the fifth dimension.

Kaluza also gave an answer as to where the fifth dimension was. Since we see no evidence of a fifth dimension, it must have "curled up" so small that it cannot be observed. (Imagine taking a two-dimensional sheet of paper and rolling it up tightly into a cylinder. From a distance, the cylinder looks like a one-dimensional line. In this way, a two-dimensional object has been turned into a one-dimensional object by curling it up.)

Kaluza's paper initially created a sensation. But in the coming years, objections were found to his theory. What was the size of this new fifth dimension? How did it curl up? No answers could be found.

For decades Einstein would work on this theory in fits and starts. After he passed away in 1955, the theory was soon forgotten, becoming just a strange footnote to the evolution of physics.

String Theory

All this has changed with the coming of a startling new theory, called the superstring theory. By the 1980s physicists were drowning in a sea of subatomic particles. Every time they smashed an atom apart with powerful particle accelerators, they found scores of new particles spitting out. It was so frustrating that J. Robert Oppenheimer declared that the Nobel Prize in Physics should go to the physicist who did *not* discover a new particle that year! (Enrico Fermi, horrified at the proliferation of subatomic particles with Greek-sounding names, said, "If I could remember the names of all these particles, I would have become a botanist.") After decades of hard work, this zoo of particles could be arranged into something called the Standard Model. Billions of dollars, the sweat of thousands of engineers and physicists, and twenty Nobel Prizes have gone into painfully assembling, piece by piece, the Standard Model. It is a truly remarkable theory, which seems to fit all the experimental data concerning subatomic physics.

But the Standard Model, for all its experimental successes, suffered from one serious defect. As Stephen Hawking says, "It is ugly and ad hoc." It contains at least nineteen free parameters (including the particle masses and the strength of their interactions with other particles), thirty-six quarks and antiquarks, three exact and redundant copies of sub-particles, and a host of strange-sounding subatomic particles, such as tau neutrinos, Yang-Mills gluons, Higgs bosons, W bosons, and Z particles. Worse, the Standard Model makes no mention of gravity. It seemed hard to believe that nature, at its most supreme, fundamental level, could be so haphazard and supremely inelegant. Here was a theory only a mother could love. The sheer inelegance of the Standard Model forced physicists to reanalyze all their assumptions about nature. Something was terribly wrong.

If one analyzes the last few centuries in physics, one of the most important achievements of the last century was to summarize all fundamental physics into two great theories: the quantum theory (represented by the Standard Model) and Einstein's theory of general relativity (describing gravity). Remarkably, together they represent the sum total of all physical knowledge at a fundamental level. The first theory describes the world of the very small, the subatomic quantum world where particles perform a fantastic dance, darting in and out of existence and appearing two places at the same time. The second theory describes the world of the very large, such as black holes and the big bang, and uses the language of smooth surfaces, stretched fabrics, and warped surfaces. The theories are opposites in every way, using different mathematics, different assumptions, and different physical pictures. It's as if nature had two hands, neither of which communicated with the other. Furthermore, any attempt to join these two theories has led to meaningless answers. For half a century any physicist who tried to mediate a shotgun wedding between the quantum theory and general relativity found that the theory blew up in their faces, producing infinite answers that made no sense.

All of this changed with the advent of the superstring theory, which posits that the electron and the other subatomic particles are nothing but different vibrations of a string, acting like a tiny rubber band. If one strikes the rubber band, it vibrates in different modes, with each note corresponding to a different subatomic particle. In this way, superstring theory explains the hundreds of subatomic particles that have been discovered so far in our particle accelerators. Einstein's theory, in fact, emerges as just one of the lowest vibrations of the string.

String theory has been hailed as a "theory of everything," the fabled theory that eluded Einstein for the last thirty years of his life. Einstein wanted a single, comprehensive theory that would summarize all physical law, that would allow him to "read the Mind of God." If string theory is correct in unifying gravity with the quantum theory, then it might represent the crowning achievement of science going back two thousand years ago to when the Greeks asked what matter was made of.

But the bizarre feature of superstring theory is that these strings can only vibrate in a specific dimension of space-time; they can only vibrate in ten dimensions. If one tries to create a string theory in other dimensions, the theory breaks down mathematically.

Our universe, of course, is four-dimensional (with three dimensions of space and one of time). This means that the other six dimensions must have collapsed somehow, or curled up, like Kaluza's fifth dimension.

Recently physicists have given serious thought to proving or disproving the existence of these higher dimensions. Perhaps the simplest way to prove the existence of higher dimensions would be to find deviations from Newton's law of gravity. In high school we learn that the gravity of the Earth diminishes as we go into outer space. More precisely, gravity diminishes with the square of the distance of separation. But this is only because we live in a three-dimensional world. (Think of a sphere surrounding the Earth. The gravity of the Earth spreads out evenly across the surface of the sphere, so the larger the sphere, the weaker the gravity. But since the surface of the sphere grows as the square of its radius, the strength of gravity, spread out over the surface of the sphere, must diminish as the square of the radius.)

But if the universe had four spatial dimensions, then gravity should diminish as the cube of the distance of separation. If the universe had n spatial dimensions, then gravity should diminish as the $n - 1$-th power. Newton's famous inverse-square law has been tested with great accuracy over astronomical distances; that is why we can send space probes soaring past the rings of Saturn with breathtaking accuracy. But until recently Newton's inverse-square law had never been tested at small distances in the laboratory.

The first experiment to test the inverse-square law at small distances was performed at the University of Colorado in 2003 with negative results. Apparently there is no parallel universe, at least not in Colorado. But this negative result has only whetted the appetite of other physicists, who hope to duplicate this experiment with even greater accuracy.

Furthermore, the Large Hadron Collider, which will become operational in 2008 outside Geneva, Switzerland, will be looking for a new type of particle called the "sparticle," or superparticle, which is a higher vibration of the superstring (everything you see around you is but the lowest vibration of the superstring). If sparticles are found by the LHC, it could signal a revolution in the way we view the universe. In this picture of the universe, the Standard Model simply represents the lowest vibration of the superstring.

Kip Thorne says, "By 2020, physicists will understand the laws of quantum gravity, which will be found to be a variant of string theory."

In addition to higher dimensions, there is another parallel universe predicted by string theory, and this is the "multiverse."

THE MULTIVERSE

There is still one nagging question about string theory: why should there be five different versions of string theory? String theory could successfully unify the quantum theory with gravity, but there were five ways in which this could be done. This was rather embarrassing, since most physicists wanted a unique "theory of everything." Einstein, for example, wanted to know if "God had any choice in making the universe." His belief was that the unified field theory of everything should be unique. So why should there be five string theories?

In 1994 another bombshell was dropped. Edward Witten of Princeton's Institute for Advanced Study and Paul Townsend of Cambridge University speculated that all five string theories were in fact the same theory–but only if we add an eleventh dimension. From the vantage point of the eleventh dimension, all five different theories collapsed into one! The theory was unique after all, but only if we ascended to the mountaintop of the eleventh dimension.

In the eleventh dimension a new mathematical object can exist, called the membrane (e.g., like the surface of a sphere). Here was the amazing observation: if one dropped from eleven dimensions down to ten dimensions, all five string theories would emerge, starting from a single membrane. Hence all five string theories were just different ways of moving a membrane down from eleven to ten dimensions.

(To visualize this, imagine a beach ball with a rubber band stretched around the equator. Imagine taking a pair of scissors and cutting the beach ball twice, once above and once below the rubber band, thereby lopping off the top and bottom of the beach ball. All that is left is the rubber band, a string. In the same way, if we curl up the eleventh dimension, all that is left of a membrane is its equator, which is the string. In fact, mathematically there are five ways in which this slicing can occur, leaving us with five different string theories in ten dimensions.)

The eleventh dimension gave us a new picture. It also meant that perhaps the universe itself was a membrane, floating in an eleven-dimensional space-time. Moreover, not all these dimensions had to be small. In fact, some of these dimensions might actually be infinite.

This raises the possibility that our universe exists in a multiverse of other universes. Think of a vast collection of floating soap bubbles or membranes. Each soap bubble represents an entire universe floating in a larger arena of eleven-dimensional hyperspace. These bubbles can join with other bubbles, or split apart, and even pop into existence and disappear. We might live on the skin of just one of these bubble universes.

Max Tegmark of MIT believes that in fifty years "the existence of these 'parallel universes' will be no more controversial than the existence of other galaxies–then called 'island universes'–was 100 years ago."

How many universes does string theory predict? One embarrassing feature of string theory is that there are trillions upon trillions of possible universes, each one compatible with relativity and the quantum theory. One estimate claims that there might be a googol of such universes. (A googol is 1 followed by 100 zeros.)

Normally communication between these universes is impossible. The atoms of our body are like flies trapped on flypaper. We can move freely about in three dimensions along our membrane universe, but we cannot leap off the universe into hyperspace, because we are glued onto our universe. But gravity, being the warping of space-time, can freely float into the space between universes.

In fact, there is one theory that states that dark matter, an invisible form of matter that surrounds the galaxy, might be ordinary matter floating in a parallel universe. As in H. G. Wells's novel *The Invisible Man,* a person would become invisible if he floated just above us in the fourth dimension. Imagine two parallel sheets of paper, with someone floating on one sheet, just above the other.

In the same way there is speculation that dark matter might be an ordinary galaxy hovering above us in another membrane universe. We could feel the gravity of this galaxy, since gravity can ooze its way between universes, but the other galaxy would be invisible to us because light moves underneath the galaxy. In this way, the galaxy would have gravity but would be invisible, which fits the description of dark matter. (Yet another possibility is that dark matter might consist of the next vibration of the superstring. Everything we see around us, such as atoms and light, is nothing but the lowest vibration of the superstring. Dark matter might be the next higher set of vibrations.)

To be sure, most of these parallel universes are probably dead ones, consisting of a formless gas of subatomic particles, such as electrons and neutrinos. In these universes the proton might be unstable, so all matter as we know it would slowly decay and dissolve. Complex matter, consisting of atoms and molecules, probably would not be possible in many of these universes.

Other parallel universes might be just the opposite, with complex forms of matter far beyond anything we can conceive of. Instead of just one type of atom consisting of protons, neutrons, and electrons, they might have a dazzling array of other types of stable matter.

These membrane universes might also collide, creating cosmic fireworks. Some physicists at Princeton believe that perhaps our universe started out as two gigantic membranes that collided 13.7 billion years ago. The shock waves from that cataclysmic collision created our universe, they believe. Remarkably, when the experimental consequences of this strange idea are explored they apparently match the results from the WMAP satellite currently orbiting the Earth. (This is called the "Big Splat" theory.)

The theory of the multiverse has one fact in its favor. When we analyze the constants of nature, we find that they are "tuned" very precisely to allow for life. If we increase the strength of the nuclear force, then the stars burn out too quickly to give rise to life. If we decrease the strength of the nuclear force, then stars never ignite at all and life cannot exist. If we increase the force of gravity, then our universe dies quickly in a Big Crunch. If we decrease the strength of gravity, then the universe expands rapidly into a Big Freeze. In fact, there are scores of "accidents" involving the constants of nature that allow for life. Apparently, our universe lives in a "Goldilocks zone" of many parameters, all of which are "fine-tuned" to allow for life. So either we are left with the conclusion that there is a God of some sort who has chosen our universe to be "just right" to allow for life, or there are billions of parallel universes, many of them dead. As Freeman Dyson has said, "The universe seemed to know we were coming."

Sir Martin Rees of Cambridge University has written that this fine tuning is, in fact, convincing evidence for the multiverse. There are five physical constants (such as the strength of the various forces) that are fine-tuned to allow for life, and he believes that there are also an infinite number of universes in which the constants of nature are not compatible with life.

This is the so-called "anthropic principle." The weak version simply states that our universe is fine-tuned to allow for life (because we are here to make this statement in the first place). The strong version says that perhaps our existence was a by-product of design or purpose. Most cosmologists would agree to the weak version of the anthropic principle, but there is considerable debate over whether the anthropic principle is a new principle of science that could lead to new discoveries and results, or whether it is simply a statement of the obvious.

Quantum Theory

In addition to higher dimensions and the multiverse, there is yet another type of parallel universe, one that gave Einstein headaches and one that continues to bedevil physicists today. This is the quantum universe predicted by ordinary quantum mechanics. The paradoxes within quantum physics seem so intractable that Nobel laureate Richard Feynman was fond of saying that *no one* really understands the quantum theory.

Ironically, although the quantum theory is the most successful theory ever proposed by the human mind (often accurate to within one part in 10 billion), it is built on a sand of chance, luck, and probabilities. Unlike Newtonian theory, which gave definite, hard answers to the motion of objects, the quantum theory can give only probabilities.

The wonders of the modern age, such as lasers, the Internet, computers, TV, cell phones, radar, microwave ovens, and so forth, are all based on the shifting sands of probabilities.

The sharpest example of this conundrum is the famous "Schrödinger's cat" problem (formulated by one of the founders of the quantum theory, who paradoxically proposed the problem in order to smash this probabilistic interpretation). Schrödinger railed against this interpretation of his theory, stating, "If one has to stick to this damned quantum jumping, then I regret having ever been involved in this thing."

The Schrödinger's cat paradox is as follows: a cat is placed in a sealed box. Inside a gun is pointed at the cat (and the trigger is then connected to a Geiger counter next to a piece of uranium). Normally when the uranium atom decays it sets off the Geiger counter and then the gun and the cat is killed. The uranium atom can either decay or not. The cat is either dead or alive. This is just common sense.

But in the quantum theory, we don't know for sure if the uranium has decayed. So we have to add the two possibilities, adding the wave function of a decayed atom with the wave function of an intact atom. But this means that, in order to describe the cat, we have to add the two states of the cat. So the cat is neither dead nor alive. It is represented as the sum of a dead cat and a live cat!

As Feynman once wrote, quantum mechanics "describes nature as absurd from the point of view of common sense. And it fully agrees with experiment. So I hope you can accept nature as She is–absurd."

To Einstein and Schrödinger, this was preposterous. Einstein believed in "objective reality," a commonsense, Newtonian view in which objects existed in definite states, not as the sum of many possible states. And yet this bizarre interpretation lies at the heart of modern civilization. Without it modern electronics (and the very atoms of our body) would cease to exist. (In our ordinary world we sometimes joke that it's impossible to be "a little bit pregnant." But in the quantum world, it's even worse. We exist simultaneously as the sum of all possible bodily states: unpregnant, pregnant, a child, an elderly woman, a teenager, a career woman, etc.)

There are several ways to resolve this sticky paradox. The founders of the quantum theory believed in the Copenhagen School, which said that once you open the box, you make a measurement and can determine if the cat is dead or alive. The wave function has "collapsed" into a single state and common sense takes over. The waves have disappeared, leaving only particles. This means that the cat now enters a definite state (either dead or alive) and is no longer described by a wave function.

Thus there is an invisible barrier separating the bizarre world of the atom and the macroscopic world of humans. For the atomic world, everything is described by waves of probability, in which atoms can be in many places at the same time. The larger the wave at some location, the greater the probability of finding the particle at that point. But for large objects these waves have collapsed and objects exist in definite states, and hence common sense prevails.

(When guests would come to Einstein's house, he would point to the moon and ask, "Does the moon exist because a mouse looks at it?" In some sense, the answer of the Copenhagen School might be yes.)

Most Ph.D. physics textbooks religiously adhere to the original Copenhagen School, but many research physicists have abandoned it. We now have nanotechnology and can manipulate individual atoms, so atoms that dart in and out of existence can be manipulated at will, using our scanning tunneling microscopes. There is no invisible "wall" separating the microscopic and macroscopic world. There is a continuum.

At present there is no consensus on how to resolve this issue, which strikes at the very heart of modern physics. At conferences, many theories heatedly compete with others. One minority point of view is that there must be a "cosmic consciousness" pervading the universe. Objects spring into being when measurements are made, and measurements are made by conscious beings. Hence there must be cosmic consciousness that pervades the universe determining which state we are in. Some, like Nobel laureate Eugene Wigner, have argued that this proves the existence of God or some cosmic consciousness. (Wigner wrote, "It was not possible to formulate the laws [of the quantum theory] in a fully consistent way without reference to consciousness." In fact, he even expressed an interest in the Vedanta philosophy of Hinduism, in which the universe is pervaded by an all-embracing consciousness.)

Another viewpoint on the paradox is the "many worlds" idea, proposed by Hugh Everett in 1957, which states that the universe simply splits in half, with a live cat in one half and a dead cat in the other. This means that there is a vast proliferation or branching of parallel universes each time a quantum event occurs. Any universe that can exist, does. The more bizarre the universe, the less likely it is, but nonetheless these universes exist. This means there is a parallel world in which the Nazis won World War II, or a world where the Spanish Armada was never defeated and everyone is speaking in Spanish. In other words, the wave function never collapses. It simply continues on its way, merrily splitting off into countless universes.

As MIT physicist Alan Guth has said, "There is a universe where Elvis is still alive, and Al Gore is President." Nobel laureate Frank Wilczek says, "We are haunted by the awareness that infinitely many slightly variant copies of ourselves are living out their parallel lives and that every moment more duplicates spring into existence and take up our many alternative futures."

One point of view that is gaining in popularity among physicists is something called "decoherence." This theory states that all these parallel universes are possibilities, but our wave function has decohered from them (i.e., it no longer vibrates in unison with them) and hence no longer interacts with them. This means that inside your living room you coexist simultaneously with the wave function of dinosaurs, aliens, pirates, unicorns, all of them believing firmly that their universe is the "real" one, but we are no longer "in tune" with them.

According to Nobel laureate Steve Weinberg, this is like tuning into a radio station in your living room. You know that your living room is flooded with signals from scores of radio stations from around the country and the world. But your radio tunes into only one station. It has "decohered" from all the other stations. (In summing up, Weinberg notes that the "many worlds" idea is "a miserable idea, except for all the other ideas.")

So does there exist the wave function of an evil Federation of Planets that plunders weaker planets and slaughters its enemies? Perhaps, but if so, we have decohered from that universe.

Quantum Universes

When Hugh Everett discussed his "many worlds" theory with other physicists, he received puzzled or indifferent reactions. One physicist, Bryce DeWitt of the University of Texas, objected to the theory because "I just can't feel myself split." But this, Everett said, is similar to the way Galileo answered his critics who said that they could not feel the Earth moving. (Eventually DeWitt was won over to Everett's side and became a leading proponent of the theory.)

For decades the "many worlds" theory languished in obscurity. It was simply too fantastic to be true. John Wheeler, Everett's adviser at Princeton, finally concluded that there was too much "excess baggage" associated with the theory. But one reason that Everett's theory is suddenly in vogue right now is because physicists are attempting to apply the quantum theory to the last domain that has resisted being quantized: the universe itself. Applying the uncertainty principle to the entire universe naturally leads to a multiverse.

The concept of "quantum cosmology" at first seems like a contradiction in terms: the quantum theory refers to the infinitesimally tiny world of the atom, while cosmology refers to the entire universe. But consider this: at the instant of the big bang, the universe was much smaller than an electron. Every physicist agrees that electrons must be quantized; that is, they are described by a probabilistic wave equation (the Dirac equation) and can exist in parallel states. Hence if electrons must be quantized, and if the universe was once smaller than an electron, then the universe must also exist in parallel states–a theory that naturally leads to a "many worlds" approach.

The Copenhagen interpretation of Niels Bohr, however, encounters problems when applied to the entire universe. The Copenhagen interpretation, although it is taught in every Ph.D.-level quantum mechanics course on Earth, depends on an "observer" making an observation and collapsing the wave function. The observation process is absolutely essential in defining the macroscopic world. But how can one be "outside" the universe while observing the entire universe? If a wave function describes the universe, then how can an "outside" observer collapse the wave function of the universe? In fact, some see the inability to observe the universe from "outside" the universe as a fatal flaw of the Copenhagen interpretation.

In the "many worlds" approach the solution to this problem is simple: the universe simply exists in many parallel states, all defined by a master wave function, called the "wave function of the universe." In quantum cosmology the universe started out as a quantum fluctuation of the vacuum, that is, as a tiny bubble in the space-time foam. Most baby universes in the space-time foam have a big bang and then immediately have a Big Crunch afterward. That is why we never see them, because they are extremely small and short-lived, dancing in and out of the vacuum. This means that even "nothing" is boiling with baby universes popping in and out of existence, but on a scale that is too small to detect with our instruments. But for some reason, one of the bubbles in the space-time foam did not recollapse into a Big Crunch, but kept on expanding. This is our universe. According to Alan Guth, this means that the entire universe is a free lunch.

In quantum cosmology, physicists start with an analogue of the Schrödinger equation, which governs the wave function of electrons and atoms. They use the DeWitt-Wheeler equation, which acts on the "wave function of the universe." Usually the Schrödinger wave function is defined at every point in space and time, and hence you can calculate the chances of finding an electron at that point in space and time. But the "wave function of the universe" is defined over all possible universes. If the wave function of the universe happens to be large when defined for a specific universe, it means that there is a good chance that the universe will be in that particular state.

Hawking has been pushing this point of view. Our universe, he claims, is special among other universes. The wave function of the universe is large for our universe and is nearly zero for most other universes. Thus there is a small but finite probability that other universes can exist in the multiverse, but ours has the largest probability. Hawking, in fact, tries to derive inflation in this way. In this picture a universe that inflates is simply more likely than a universe that does not, and hence our universe has inflated.

The theory that our universe came from the "nothingness" of the space-time foam might seem to be totally untestable, but it is consistent with several simple observations. First, many physicists have pointed out that it is astonishing that the total amount of positive charges and negative charges in our universe comes out to exactly zero, at least to within experimental accuracy. We take it for granted that in outer space gravity is the dominant force, yet this is only because the positive and negative charges cancel out precisely. If there was the slightest imbalance between positive and negative charges on the Earth, it might be sufficient to rip the Earth apart, overcoming the gravitational force that holds the Earth together. One simple way to explain why there is this balance between positive and negative charges is to assume that our universe came from "nothing," and "nothing" has zero charge.

Second, our universe has zero spin. Although for years Kurt Gödel tried to show that the universe was spinning by adding up the spins of the various galaxies, astronomers today believe that the total spin of the universe is zero. The phenomenon would be easily explained if the universe came from "nothing," since "nothing" has zero spin.

Third, our universe's coming from nothing would help to explain why the total matter-energy content of the universe is so small, perhaps even zero. When we add up the positive energy of matter and the negative energy associated with gravity, the two seem to cancel each other out. According to general relativity, if the universe is closed and finite, then the total amount of matter-energy in the universe should be exactly zero. (If our universe is open and infinite, this does not have to be true, but inflationary theory does seem to indicate that the total amount of matter-energy in our universe is remarkably small.)

Contact Between Universes?

This leaves open some tantalizing questions: If physicists can't rule out the possibility of several types of parallel universes, would it be possible to make contact with them? To visit them? Or is it possible that perhaps beings from other universes have visited us?

Contact with other quantum universes that have decohered from us seems highly unlikely. The reason that we have decohered from these other universes is that our atoms have bumped into countless other atoms in the surrounding environment. Each time a collision occurs, the wave function of that atom appears to "collapse" a bit; that is, the number of parallel universes decreases. Each collision narrows the number of possibilities. The sum total of all these trillions of atomic "mini-collapses" gives the illusion that the atoms of our body are totally collapsed in a definite state. The "objective reality" of Einstein is an illusion created by the fact that we have so many atoms in our body, each one bumping into others, each time narrowing the number of possible universes.

It's like looking at an out-of-focus image through a camera. This would correspond to the microworld, where everything seems fuzzy and indefinite. But each time you adjust the focus of the camera, the image gets sharper and sharper. This corresponds to trillions of tiny collisions with neighboring atoms, each of which reduces the number of possible universes. In this way, we smoothly make the transition from the fuzzy microworld to the macroworld.

So the probability of interacting with another quantum universe similar to ours is not zero, but it decreases rapidly with the number of atoms in your body. Since there are trillions upon trillions of atoms in your body, the chance that you will interact with another universe consisting of dinosaurs or aliens is infinitesimally small. You can calculate that you would have to wait much longer than the lifetime of the universe for such an event to happen.

So contact with a quantum parallel universe cannot be ruled out, but it would be an exceedingly rare event since we have decohered from them. But in cosmology, we encounter a different type of parallel universe: a multiverse of universes that coexist with each other, like soap bubbles floating in a bubble bath. Contact with another universe in the multiverse is a different question. It would undoubtedly be a difficult feat, but one that might be possible for a Type III civilization.

As we discussed before, the energy necessary to open a hole in space or to magnify the space-time foam is on the order of the Planck energy, where all known physics breaks down. Space and time are not stable at that energy, and this opens the possibility of leaving our universe (assuming that other universes exist and we are not killed in the process).

This is not a purely academic question, since all intelligent life in the universe will one day have to confront the end of the universe. Ultimately, the theory of the multiverse may be the salvation for all intelligent life in our universe. Recent data from the WMAP satellite currently orbiting the Earth confirms that the universe is expanding at an accelerating rate. One day we may all perish in what physicists refer to as the Big Freeze. Eventually, the entire universe will go black; all the stars in the heavens will blink out and the universe will consist of dead stars, neutron stars, and black holes. Even the very atoms of their bodies may begin to decay. Temperatures may plunge to near absolute zero, making life impossible.

As the universe approaches that point, an advanced civilization facing the ultimate death of the universe could contemplate taking the ultimate journey to another universe. For these beings the choice would be to freeze to death or leave. The laws of physics are a death warrant for all intelligent life, but there is an escape clause in those laws.

Such a civilization would have to harness the power of huge atom smashers and laser beams as large as a solar system or star cluster to concentrate enormous power at a single point in order to attain the fabled Planck energy. It is possible that doing so would be sufficient to open up a wormhole or gateway to another universe. A Type III civilization may use the colossal energy at their disposal to open a wormhole as it makes a journey to another universe, leaving our dying universe and starting over again.

A Baby Universe in the Laboratory?

As far-fetched as some of these ideas appear, they have been seriously considered by physicists. For example, when trying to understand how the big bang got started, we have to analyze the conditions that may have led to that original explosion. In other words, we have to ask: how do you make a baby universe in the laboratory? Andrei Linde of Stanford University, one of the cocreators of the inflationary universe idea, says that if we can create baby universes, then "maybe it's time we redefine God as something more sophisticated than just the creator of the universe."

The idea is not new. Years ago when physicists calculated the energy necessary to ignite the big bang "people immediately started to wonder what would happen if you put lots of energy in one space in the lab–shot lots of cannons together. Could you concentrate enough energy to set off a mini big bang?" asks Linde.

If you concentrated enough energy at a single point all you would get would be a collapse of space-time into a black hole, nothing more. But in 1981 Alan Guth of MIT and Linde proposed the "inflationary universe" theory, which has since generated enormous interest among cosmologists. According to this idea, the big bang started off with a turbocharged expansion, much faster than previously believed. (The inflationary universe idea solved many stubborn problems in cosmology, such as why the universe should be so uniform. Everywhere we look, from one part of the night sky to the opposite side, we see a uniform universe, even though there has not been enough time since the big bang for these vastly separated regions to be in contact. The answer to this puzzle, according to the inflationary universe theory, is that a tiny piece of space-time that was relatively uniform blew up to become the entire visible universe.) In order to jump-start inflation, Guth assumed that at the beginning of time there were tiny bubbles of space-time, one of which inflated enormously to become the universe of today.

In one swoop the inflationary universe theory answered a host of cosmological questions. Moreover, it is consistent with all the data pouring in today from outer space from the WMAP and COBE satellites. It is, in fact, unquestionably the leading candidate for a theory of the big bang.

Yet the inflationary universe theory raises a series of embarrassing questions. Why did this bubble start to inflate? What turned off the expansion, resulting in the present-day universe? If inflation happened once, could it happen again? Ironically, although the inflation scenario is the leading theory in cosmology, almost nothing is known about what set the inflation into motion and why it stopped.

In order to answer these nagging questions, in 1987 Alan Guth and Edward Fahri of MIT asked another hypothetical question: how might an advanced civilization inflate its own universe? They believed that if they could answer this question, they might be able to answer the deeper question of why the universe inflated to begin with.

They found that if you concentrated enough energy at a single point, tiny bubbles of space-time would form spontaneously. But if the bubbles were too small, they would disappear back into the space-time foam. Only if the bubbles were big enough could they expand into an entire universe.

On the outside the birth of this new universe would not look like much, perhaps no more than the detonation of a 500-kiloton nuclear bomb. It would appear as if a small bubble had disappeared from the universe, leaving a small nuclear explosion. But inside the bubble an entirely new universe might expand out. Think of a soap bubble that splits or buds a smaller bubble, creating a baby soap bubble. The tiny soap bubble might expand rapidly into an entirely new soap bubble. Likewise, inside the universe you would see an enormous explosion of space-time and the creation of an entire universe.

Since 1987 many theories have been proposed to see if the introduction of energy can make a large bubble expand into an entire universe. The most commonly accepted theory is that a new particle, called the "inflaton," destabilized space-time, causing these bubbles to form and expand.

The latest controversy erupted in 2006 when physicists began to look seriously at a new proposal to ignite a baby universe with a monopole. Although monopoles–particles that carry only a single north or south pole–have never been seen, it is believed that they dominated the original early universe. They are so massive that they are extremely hard to create in the laboratory, but precisely because they are so massive, if we injected even more energy into a monopole we might be able to ignite a baby universe into expanding into a real universe.

Why would physicists want to create a universe? Linde says, "In this perspective, each of us can become a god." But there is a more practical reason for wanting to create a new universe: ultimately, to escape the eventual death of our universe.

The Evolution of Universes?

Some physicists have taken this idea even further, to the very limits of science fiction, in asking whether intelligence may have had a hand in designing our universe.

In the Guth/Fahri picture, an advanced civilization can create a baby universe, but the physical constants (e.g., the mass of the electron and proton and the strengths of the four forces) are the same. But what if an advanced civilization could create baby universes that differ slightly in their fundamental constants? Then the baby universes would be able to "evolve" with time, with each generation of baby universes being slightly different from the previous generation.

If we consider the fundamental constants to be the "DNA" of a universe, it means that intelligent life might be able to create baby universes with slightly different DNA. Eventually, universes would evolve, and the universes that proliferated would be those that had the best "DNA" that allow for the flourishing of intelligent life. Physicist Edward Harrison, building on a previous idea by Lee Smolin, has proposed a "natural selection" among universes. The universes that dominate the multiverse are precisely those that have the best DNA, which is compatible with creating advanced civilizations, which in turn create more baby universes. "Survival of the fittest" simply means survival of the universes that are most favorable to producing advanced civilizations.

If this picture is correct, it would explain why the fundamental constants of the universe are "fine-tuned" to allow for life. It simply means that universes with desirable fundamental constants compatible with life are the ones that proliferate in the multiverse.

(Although this "evolution of universes" idea is attractive because it might be able to explain the anthropic principle problem, the difficulty with this idea is that it is untestable and unfalsifiable. We will have to wait until we have a complete theory of everything before we can make sense out of this idea.)

Currently, our technology is far too primitive to reveal the presence of these parallel universes. So all this would qualify as a Class II impossibility–impossible today, but not in violation of the laws of physics. On a scale of thousands to millions of years, these speculations could become the basis of a new technology for a Type III civilization.

Part III

CLASS III IMPOSSIBILITIES

14: PERPETUAL MOTION MACHINES

> Theories have four stages of acceptance:
>
> i. this is worthless nonsense;
>
> ii. this is interesting, but perverse;
>
> iii. this is true, but quite unimportant;
>
> iv. I always said so.
>
> –J. B. S. HALDANE, 1963

In Isaac Asimov's classic novel *The Gods Themselves,* an obscure chemist in the year 2070 accidentally stumbles upon the greatest discovery of all time, the Electron Pump, which produces unlimited energy for free. The impact is immediate and profound. He is hailed as the greatest scientist of all time for satisfying civilization's unquenchable thirst for energy. "It was Santa Claus and Aladdin's lamp of the whole world," Asimov wrote. The company he forms soon becomes one of the richest corporations on the planet, putting the oil, gas, coal, and nuclear industries out of business.

The world is awash with free energy and civilization is drunk with this newfound power. As everyone celebrates this great achievement, one lone physicist is uneasy. "Where is all this free energy coming from?" he asks himself. Eventually he uncovers the secret. The free energy comes with a terrible price. This energy is pouring in from a hole in space connecting our universe to a parallel universe, and the sudden influx of energy into our universe is setting off a chain reaction that will eventually destroy the stars and galaxies, turning the sun into a supernova, and destroying the Earth with it.

Since recorded history, the holy grail for inventors, scientists, as well as charlatans and scam artists has been the fabled "perpetual motion machine," a device that runs forever without any loss of energy. An even better version is a device that can create *more* energy than it consumes, such as the Electron Pump, which creates free, limitless energy.

In the coming years, as our industrialized world gradually runs out of cheap oil, there will be enormous pressure to find abundant new sources of clean energy. Soaring gas prices, falling production, increased pollution, atmospheric changes–all are fueling a renewed, intense interest in energy.

Today a few inventors riding this wave of concern promise to deliver unlimited quantities of free energy, offering to sell their inventions for hundreds of millions. Scores of investors periodically line up, lured by sensational claims in the financial media that often hail these mavericks as the next Edison.

The popularity of perpetual motion machines is widespread. On an episode of *The Simpsons,* entitled "The PTA Disbands," Lisa builds her own perpetual motion machine during a teachers' strike. This prompts Homer to declare sternly, "Lisa, get in here . . . in this house we obey the laws of thermodynamics!"

In the computer games *The Sims, Xenosaga Episodes I* and *II,* and *Ultima VI: The False Prophet,* as well as the Nickelodeon program *Invader Zim,* perpetual motion machines figure prominently in the plots.

But if energy is so precious, then precisely what is the likelihood of our creating a perpetual motion machine? Are these devices truly impossible, or would their creation require a revision in the laws of physics?

History Viewed Through Energy

Energy is vital to civilization. In fact, all of human history can be viewed through the lens of energy. For 99.9 percent of human existence, primitive societies were nomadic, scratching a meager living scavenging and hunting for food. Life was brutal and short. The energy available to us was one-fifth of a horsepower–the power of our own muscles. Analyses of the bones of our ancestors show evidence of enormous wear and tear, caused by the crushing burdens of daily survival. Average life expectancy was less than twenty years.

But after the end of the last ice age about ten thousand years ago, we discovered agriculture and domesticated animals, especially the horse, gradually raising our energy output to one or two horsepower. This set into motion the first great revolution in human history. With the horse or ox, one man had enough energy to plow an entire field by himself, travel tens of miles in a day, or move hundreds of pounds of rock or grain from one place to another. For the first time in human history, families had a surplus of energy, and the result was the founding of our first cities. Excess energy meant that society could afford to support a class of artisans, architects, builders, and scribes, and thus ancient civilization could flourish. Soon great pyramids and empires rose from the jungles and desert. Average life expectancy reached about thirty years.

Then about three hundred years ago the second great revolution in human history took place. With the coming of machines and steam power, the energy available to a single person soared to tens of horsepower. By harnessing the power of the steam locomotive, people could now cross entire continents in a few days. Machines could plow entire fields, transport hundreds of passengers thousands of miles, and allow us to build huge towering cities. Average life expectancy by 1900 had reached almost fifty in the United States.

Today we are in the midst of the third great revolution in human history, the information revolution. Because of an exploding population and our ravenous appetite for electricity and power, our energy needs have skyrocketed and our energy supply is being stretched to the very limit. The energy available to a single individual is now measured in thousands of horsepower. We take for granted that a single car can generate hundreds of horsepower. Not surprisingly, this demand for more and more energy has sparked an interest in greater sources of energy, including perpetual motion machines.

Perpetual Motion Machines Through History

The search for a perpetual motion machine is an ancient one. The first recorded attempt to build a perpetual motion machine dates back to the eighth century in Bavaria. It was a prototype for hundreds of variations to come for the next thousand years; it was based on a series of small magnets attached to a wheel, like a Ferris wheel. The wheel was placed on top of a much larger magnet on the floor. As each magnet on the wheel passed over the stationary magnet, it was supposed to be attracted then repelled by the larger magnet, thereby pushing the wheel and creating perpetual motion.

Another ingenious design was devised in 1150 by the Indian philosopher Bhaskara, who proposed a wheel that would run forever by adding a weight to the rim, causing the wheel to spin because it was unbalanced. Work would be done by the weight as it made a revolution, and then it would return to its original position. By iterating this over and over again, Bhaskara claimed that one could extract unlimited work for free.

The Bavarian and the Bhaskara designs for perpetual motion machines and their many descendants all share the same ingredients: a wheel of some sort that can make a single revolution without the addition of any energy, producing usable work in the process. (Careful examination of these ingenious machines usually shows that energy is actually lost in each cycle, or that no usable work can be extracted.)

The coming of the Renaissance accelerated proposals for a perpetual motion machine. In 1635 the first patent was granted for a perpetual motion machine. By 1712 Johann Bessler had analyzed some three hundred different models and proposed a design of his own. (According to legend, his maid later exposed his machine as a fraud.) Even the great renaissance painter and scientist Leonardo da Vinci became interested in perpetual motion machines. Although he denounced them in public, comparing them to the fruitless search for the philosopher's stone, in private he made ingenious sketches in his notebooks of self-propelling, perpetual motion machines, including a centrifugal pump and a chimney jack used to turn a roasting skewer over a fire.

By 1775 so many designs were being proposed that the Royal Academy of Science in Paris stated that it would "no longer accept or deal with proposals concerning perpetual motion."

Arthur Ord-Hume, a historian of these perpetual motion machines, has written about the tireless dedication of these inventors, working against incredible odds, comparing them to the ancient alchemists. But, he noted, "Even the alchemist . . . knew when he was beaten."

Hoaxes and Frauds

The incentive to produce a perpetual motion machine was so great that hoaxes became commonplace. In 1813 Charles Redheffer exhibited a machine in New York City that amazed audiences, producing unlimited energy for free. (But when Robert Fulton examined the machine carefully, he found a hidden cat-gut belt driving the machine. This cable was in turn connected to a man secretly turning a crank in the attic.)

Scientists and engineers, too, got swept up in the enthusiasm for perpetual motion machines. In 1870 the editors of *Scientific American* were fooled by a machine built by E. P. Willis. The magazine ran a story with the sensational title "Greatest Discovery Ever Yet Made." Only later did investigators discover that there was a hidden source of energy for Willis's perpetual motion machine.

In 1872 John Ernst Worrell Kelly perpetrated the most sensational and lucrative scam of his day, swindling investors of nearly $5 million, a princely sum back in the late nineteenth century. His perpetual motion machine was based on resonating tuning forks that he claimed tapped into the "ether." Kelly, a man with no scientific background, would invite wealthy investors to his house, where he would amaze them with his Hydro-Pneumatic-Pulsating-Vacuo-Engine, which whizzed around without any external power source. Eager investors, amazed by this self-propelled machine, flocked to pour money into his coffers.

Later some disillusioned investors angrily accused him of fraud, and he actually spent some time in jail, although he died a wealthy man. After his death investigators found the clever secret of his machine. When his house was torn down concealed tubes were found in the floor and walls of the basement that secretly delivered compressed air to his machines. These tubes in turn were energized by a flywheel.

Even the U.S. Navy and the president of the United States were taken in by such a machine. In 1881 John Gamgee invented a liquid ammonia machine. The vaporization of cold ammonia would create expanding gases that could move a piston, and hence could power machines using only the heat of the oceans themselves. The U.S. Navy was so enthralled with the idea of extracting unlimited energy from the oceans that it approved the device and even demonstrated it to President James Garfield. The problem was that the vapor did not condense back into a liquid properly; hence the cycle could not be completed.

So many proposals for perpetual motion machines have been presented to the U.S. Patent and Trademark Office (USPTO) that the office refuses to grant a patent for such a device unless a working model is presented. In certain rare circumstances, when the patent examiners can find nothing obviously wrong with a model, a patent is granted. The USPTO states, "With the exception of cases involving perpetual motion, a model is not ordinarily required by the Office to demonstrate the operability of a device." (This loophole has allowed unscrupulous inventors to persuade naïve investors to finance their inventions by claiming that the USPTO has officially recognized their machine.)

The pursuit of the perpetual motion machine, however, has not been fruitless from a scientific point of view. On the contrary, although inventors have never produced a perpetual motion machine, the enormous time and energy invested in building such a fabled machine has led physicists to carefully study the nature of heat engines. (In the same way, the fruitless search of alchemists for the philosopher's stone, which can turn lead into gold, helped to uncover some of the basic laws of chemistry.)

For example, in the 1760s John Cox developed a clock that could actually run forever, powered by changes in atmospheric pressure. Changes in air pressure would drive a barometer, which would then turn the hands of the clock. This clock actually worked and exists even today. The clock can run forever because energy is extracted from the outside in the form of changes in atmospheric pressure.

Perpetual motion machines like Cox's eventually led scientists to hypothesize that such machines could run forever only if energy was brought in to the device from the outside, that is, that total energy was conserved. This theory eventually led to the First Law of Thermodynamics–that the total amount of matter and energy cannot be created or destroyed. Eventually three laws of thermodynamics were postulated. The Second Law states that the total amount of entropy (disorder) always increases. (Crudely speaking, this law says that heat flows spontaneously only from hotter to colder places.) The Third Law states that you can never reach absolute zero.

If we compare the universe to a game and the goal of this game is to extract energy, then the three laws can be rephrased as follows:

"You can't get something for nothing." (First Law)
"You can't break even." (Second Law)
"You can't even get out of the game." (Third Law)

(Physicists are careful to state that these laws are not necessarily absolutely true for all time. Nevertheless, no deviation has ever been found. Anyone trying to disprove these laws must go against centuries of careful scientific experiments. We will discuss possible deviations from these laws shortly.)

These laws, among the crowning achievements of nineteenth-century science, are marked by tragedy as well as triumph. One of the key figures in formulating these laws, the great German physicist Ludwig Boltzmann, committed suicide, in part because of the controversy he created in formulating these laws.

Ludwig Boltzmann and Entropy

Boltzmann was a short, barrel-chested bear of a man, with a huge, forestlike beard. His formidable and ferocious appearance, however, belied all the wounds he suffered in defending his ideas. Although Newtonian physics was firmly established by the nineteenth century, Boltzmann knew these laws had never been rigorously applied to the controversial concept of atoms, a concept that was still not accepted by many leading scientists. (We sometimes forget that as late as a century ago there were legions of scientists who insisted that the atom was just a clever gimmick, not a real entity. Atoms were so impossibly tiny, they claimed, that they probably didn't exist at all.)

Newton showed that mechanical forces, not spirits or desires, were sufficient to determine the motion of all objects. Boltzmann then elegantly derived many of the laws of gases by a simple assumption: that gases were made of tiny atoms that, like billiard balls, obeyed the laws of forces laid down by Newton. To Boltzmann, a chamber containing gas was like a box filled with trillions of tiny steel balls, each one bouncing off the walls and each other according to Newton's laws of motion. In one of the greatest masterpieces in physics, Boltzmann (and independently James Clerk Maxwell) mathematically showed how this simple assumption could result in dazzling new laws and open up a new branch of physics called statistical mechanics.

Suddenly many of the properties of matter could be derived from first principles. Since Newton's laws stipulated that energy must be conserved when applied to atoms, each collision between atoms conserved energy; that meant that an entire chamber of trillions of atoms also conserved energy. The conservation of energy could now be established not just via experimentation, but from first principles, that is, the Newtonian motion of atoms.

But in the nineteenth century the existence of atoms was still hotly debated and often ridiculed by prominent scientists, such as philosopher Ernst Mach. A sensitive and often depressed man, Boltzmann uncomfortably found himself the lightning rod, the focus of the often vicious attacks by the anti-atomists. To the anti-atomists, anything that could not be measured did not exist, including atoms. To add to Boltzmann's humiliation, many of his papers were rejected by the editor of a prominent German physics journal because the editor insisted that atoms and molecules were strictly convenient theoretical tools, rather than objects that really existed in nature.

Exhausted and embittered from all the personal attacks, Boltzmann hung himself in 1906 while his wife and child were at the beach. Sadly he did not realize that just a year before, a brash young physicist by the name of Albert Einstein had done the impossible: he had written the first paper demonstrating the existence of atoms.

Total Entropy Always Increases

The work of Boltzmann and other physicists helped to clarify the nature of perpetual motion machines, sorting them into two types. Perpetual motion machines of the first type are those that violate the First Law of Thermodynamics; that is, they actually produce more energy than they consume. In every case physicists found that this type of perpetual motion machine relied on hidden, outside sources of energy, either through fraud, or because the inventor did not realize the source of the outside energy.

Perpetual motion machines of the second type are more subtle. They obey the First Law of Thermodynamics–conserving energy–but violate the Second Law. In theory, a perpetual motion machine of the second type produces no waste heat, so it is 100 percent efficient. Yet the Second Law says that such a machine is impossible–that waste heat must always be produced–and hence disorder or chaos in the universe, or entropy, always increases. No matter how efficient a machine might be, it will always produce some waste heat, thereby raising the entropy of the universe.

The fact that total entropy always increases lies at the heart of human history as well as mother nature. According to the Second Law, it is far easier to destroy than to build. Something that might take thousands of years to create, such as the great Aztec Empire in Mexico, can be destroyed in a matter of months; and this is what happened when a raggedy band of Spanish conquistadores, armed with horses and firearms, completely shattered that empire.

Every time you look in a mirror and see a new wrinkle or a white hair you are observing the effects of the Second Law. Biologists tell us that the aging process is the gradual accumulation of genetic errors in our cells and genes, so that the cell's ability to function slowly deteriorates. Aging, rusting, rotting, decay, disintegration, and collapse are also examples of the Second Law.

Remarking on the profound nature of the Second Law, astronomer Arthur Eddington once said, "The law that entropy always increases holds, I think, the supreme position among the laws of Nature . . . if your theory is found to be against the second law of thermodynamics, I can give you no hope; there is nothing for it but to collapse in deepest humiliation."

Even today enterprising engineers (and clever charlatans) continue to announce the invention of perpetual motion machines. Recently I was asked by the *Wall Street Journal* to comment on the work of an inventor who had actually persuaded investors to sink millions of dollars into his machine. Breathless articles were published in major financial newspapers, written by journalists with no background in science, gushing about the potential of this invention to change the world (and generate fabulous, lucrative profits in the process). "Genius or crackpot?" the headlines blared.

Investors threw enormous bundles of cash at this device, which violated the most basic laws of physics and chemistry taught in high school.

(What was shocking to me was not that a person was trying to swindle the unwary–this has been true since the dawn of time. What was surprising was that it was so easy for this inventor to fool wealthy investors because of their lack of understanding of elementary physics.) I repeated to the *Journal* the proverb "A fool and his money are easily parted" and P. T. Barnum's famous dictum "There's a sucker born every minute." Perhaps not surprisingly, the *Financial Times*, the *Economist*, and the *Wall Street Journal* have all run large feature articles on various inventors touting their perpetual motion machines.

The Three Laws and Symmetries

But all this raises a deeper question: Why do the iron laws of thermodynamics hold in the first place? It is a mystery that has intrigued scientists since the laws were first proposed. If we could answer that question, perhaps we might find loopholes in the laws, and the implications would be earth-shattering.

In graduate school I was left speechless one day when I finally learned the true origin of the conservation of energy. One of the fundamental principles of physics (discovered by mathematician Emmy Noether in 1918) is that whenever a system possesses symmetry, the result is a conservation law. If the laws of the universe remain the same over time, then the astonishing result is that the system conserves energy. (Furthermore, if the laws of physics remain the same if you move in any direction, then momentum is conserved in any direction as well. And if the laws of physics remain the same under a rotation, then angular momentum is conserved.)

This was staggering to me. I realized that when we analyze starlight from distant galaxies that are billions of light-years away, at the very edge of the visible universe, we find that the spectrum of light is identical to spectra that we can find on Earth. In the relic light that was emitted billions of years before Earth or the sun was born, we see the same unmistakable "fingerprints" of the spectrum of hydrogen, helium, carbon, neon, and so forth, that we find on the Earth today. In other words, the basic laws of physics haven't changed for billions of years, and they are constant out to the outer edges of the universe.

At a minimum, I realized, Noether's theorem means that the conservation of energy will probably last for billions of years, if not forever. As far as we know, none of the fundamental laws of physics have changed with time, and this is the reason that energy is conserved.

The implications of Noether's theorem on modern physics are profound. Whenever physicists create a new theory, whether it addresses the origin of the universe, the interactions of quarks and other subatomic particles, or antimatter, we first start with the symmetries that the system obeys. In fact, symmetries are now known to be the fundamental guiding principle in creating any new theory. In the past, symmetries were thought to be by-products of a theory–a cute but ultimately useless feature of a theory, pretty, but not essential. Today we realize that symmetries are the essential feature that defines any theory. In creating new theories, we physicists first start with symmetry, and then build the theory around it.

(Sadly, Emmy Noether, like Boltzmann before her, had to fight tooth and nail for recognition. A woman mathematician, she was denied a permanent position at leading institutions because of her sex. Noether's mentor, the great mathematician David Hilbert, was so frustrated in failing to secure a teaching appointment for Noether that he exclaimed, "What are we, a university or a bathing society?")

This raises a disturbing question. If energy is conserved because the laws of physics do not change with time, then could this symmetry be broken in rare, unusual circumstances? There is still the possibility that the conservation of energy might be violated on a cosmic scale, if the symmetry of our laws is broken in exotic and unexpected places.

One way that might happen is if the laws of physics vary with time or change with distance. (In Asimov's novel *The Gods Themselves* this symmetry was broken because there was a hole in space connecting our universe with a parallel universe. The laws of physics change in the vicinity of the hole in space, therefore allowing a breakdown in the laws of thermodynamics. Hence the conservation of energy could be violated if there are holes in space, that is, wormholes.)

Another loophole that is hotly being debated today is whether energy may spring from nothing.

ENERGY FROM THE VACUUM?

A tantalizing question is: Is it possible to extract energy from nothing? Physicists have only recently realized that the "nothing" of the vacuum is not empty at all, but teaming with activity.

One of the proponents of this idea was the eccentric genius of the twentieth century Nikola Tesla, a worthy rival to Thomas Edison. He was also one of the proponents of zero-point energy, that is, the idea that the vacuum may possess untold quantities of energy. If true, the vacuum would be the ultimate "free lunch," capable of providing unlimited energy literally from thin air. The vacuum, instead of being considered empty and devoid of any matter, would be the ultimate storehouse of energy.

Tesla was born in a small town in what is now Serbia and arrived penniless in the United States in 1884. Soon he became an assistant to Thomas Edison, but because of his brilliance, he became a rival. In a celebrated contest, which historians dubbed "The War of the Currents," Tesla was pitted against Edison. Edison believed that he could electrify the world with his direct current (DC) motors, while Tesla was the originator of alternating current (AC) and successfully showed that his methods were far superior to Edison's and incurred significantly less power loss over distance. Today the entire planet is electrified on the basis of the patents of Tesla, not Edison.

Tesla's inventions and patents number over seven hundred and contain some of the most important milestones in modern electrical history. Historians have made a credible case that Tesla invented radio before Guglielmo Marconi (widely recognized as the inventor of radio) and was working with X-rays before their official discovery by Wilhelm Roentgen. (Both Marconi and Roentgen would later win the Nobel Prize for discoveries probably made by Tesla years earlier.)

Tesla also believed that he could extract unlimited energy from the vacuum, a claim that unfortunately he did not prove in his notes. At first, "zero-point energy" (or the energy contained in a vacuum) seems to violate the First Law of Thermodynamics. Although zero-point energy defies the laws of Newtonian mechanics, the notion of zero-point energy has reemerged recently from a novel direction.

When scientists analyze the data from satellites currently orbiting the Earth, such as the WMAP satellite, they have come to the astounding conclusion that fully 73 percent of the universe is made of "dark energy," the energy of a pure vacuum. This means that the greatest reservoir of energy in the entire universe is the vacuum that separates the galaxies in the universe. (This dark energy is so colossal that it is pushing the galaxies away from each other, and may eventually rip the universe apart in a Big Freeze.)

Dark energy is everywhere in the universe, even in your living room and inside your body. The amount of dark energy in outer space is truly astronomical, outweighing all the energy of the stars and galaxies put together. We can also calculate the amount of dark energy on the Earth, and it is quite small, too small to be used to power a perpetual motion machine. Tesla was right about dark energy but wrong about the amount of dark energy on the Earth.

Or was he?

One of the most embarrassing gaps in modern physics is that no one can calculate the amount of dark energy that we can measure via our satellites. If we use the latest theory of atomic physics to calculate the amount of dark energy in the universe, we arrive at a number that is wrong by a factor of 10^{120}! That is "one" followed by 120 zeros! This is by far the largest mismatch between theory and experiment in all of physics.

The point is that no one knows how to calculate the "energy of nothing." This is one of the most important questions in physics (because it will eventually determine the fate of the universe), but at the present time we are clueless as to how to calculate it. No theory can explain dark energy, although experimental evidence for its existence is staring us in the face.

So the vacuum does have energy, as Tesla suspected. But the

amount of energy is probably too small to be used as a source of usable energy. There are vast amounts of dark energy between the galaxies, but the amount that can be found on the Earth is tiny. But the embarrassing thing is that no one knows how to calculate this energy, or where it came from.

My point is that the conservation of energy arises from deep, cosmological reasons. Any violation of these laws would necessarily mean a profound shift in our understanding of the evolution of the universe. And the mystery of dark energy is forcing physicists to confront this question head-on.

Because creating a true perpetual motion machine may require us to reevaluate the fundamental laws of physics on a cosmological scale, I would rank perpetual motion machines as a Class III impossibility; that is, either they are truly impossible, or we would need to fundamentally change our understanding of fundamental physics on a cosmological scale in order to make such a machine possible. Dark energy remains one of the great unfinished chapters in modern science.

15: PRECOGNITION

A paradox is truth standing on its head to attract attention.

—NICHOLAS FALLETTA

Is there such a thing as precognition, or seeing the future? This ancient concept is present in every religion, going back to the oracles of the Greeks and Romans and to the prophets of the Old Testament. But in such tales, the gift of prophecy can also be a curse. In Greek mythology there is the tale of Cassandra, the daughter of the King of Troy. Because of her beauty she attracted the attention of the sun god, Apollo. To win her over Apollo granted her the ability to see the future. But Cassandra spurned the advances of Apollo. In a fit of anger, Apollo twisted his gift, so that Cassandra would be able to see the future but no one would believe her. When Cassandra warned the people of Troy of their impending doom, no one listened. She foretold the treachery of the Trojan horse, the death of Agamemnon, and even her own demise, but instead of taking heed, the people of Troy thought she was mad and locked her up.

Nostradamus, writing in the sixteenth century, and more recently Edgar Cayce have claimed that they could lift the veil of time. Although there have been many claims that their predictions have come true (for example, correctly predicting World War II, JFK's assassination, and the fall of Communism), the obscure, allegorical way in which many of these seers recorded their verses allows for a variety of contradictory interpretations. The quatrains of Nostradamus, for example, are so general that one can read almost anything into them (and people have). One quatrain reads:

Earth-shaking fires from the world's center roar:
Around "New City" is the Earth a-quiver
Two nobles long shall wage a fruitless war
The nymph of springs pour forth a new, red river.

Some have claimed that this quatrain proved that Nostradamus foresaw the burning of the Twin Towers in New York on September 11, 2001. Yet over the centuries scores of other interpretations have been given to this same quatrain. The images are so vague that many interpretations are possible.

Precognition is also a favorite device of playwrights who write of the impending doom of kings and the fall of empires. In Shakespeare's *Macbeth,* precognition is central to the theme of the play and to the ambitions of Macbeth, who encounters three witches who foresee his rise to become King of Scotland. With his murderous ambitions fired up by the witches' prophesy, he begins a bloody and grisly campaign to wipe out his enemies, including killing the innocent wife and children of his rival Macduff.

After committing a series of hideous deeds to seize the crown, Macbeth learns from the witches that he cannot be defeated in battle or "vanquish'd be until great Birnam Wood to high (Dunsinane Hills) shall come against him," and that "none of woman born shall harm Macbeth." Macbeth takes comfort in this prophecy, since a forest cannot move, and all men are born of women. But the Great Birnam forest does move, as the troops of Macduff camouflaging themselves beneath twigs from the Great Birnam forest, advance on Macbeth, and Macduff himself was born via cesarean.

Although prophecies from the past have so many alternative interpretations, and hence are impossible to test, one set of prophecies is easy to analyze: predictions of the precise date of the end of the Earth–Doomsday. Ever since the last chapter of the Bible, Revelations, laid out in graphic detail the final days of the Earth, when chaos and destruction will accompany the arrival of the Antichrist and the final Second Coming of Christ, fundamentalists have tried to predict the precise date of the End of Days.

One of the most celebrated of all Doomsday predictions was made by astrologers who predicted a great flood that would end the world on February 20, 1524, based on the conjunction of all the planets in the heavens: Mercury, Venus, Mars, Jupiter, and Saturn. Mass panic swept across Europe. In England, twenty thousand people fled their homes in desperation. A fortress stocked with food and water to last two months was built around St. Bartholomew's Church. Across Germany and France, people furiously set out to build large arks to ride out the flood. Count Von Iggleheim even built a huge, three-story ark in preparation for this momentous event. But when the date finally arrived, there was only a slight rain. The mood of the crowd suddenly swung from fear into anger. People who had sold all their belongings and turned their lives upside down felt betrayed. Angry mobs began to run amok. The count was stoned to death, and hundreds were killed when the mob stampeded.

Christians aren't the only ones who feel the lure of prophecy. In 1648 Sabbatai Zevi, the son of a wealthy Jew in Smyrna, declared himself to be the Messiah and predicted that the world would end in 1666. Handsome, charismatic, and well versed in the mystical texts of the Kabbalah, he quickly assembled a group of fiercely loyal followers, who spread the news across Europe. In the spring of 1666 Jews from as far away as France, Holland, Germany, and Hungary began to pack their bags and heed the call of their Messiah. But later that year Zevi was arrested by the grand vizier in Constantinople and thrown in prison in chains. Facing a possible death sentence, he dramatically cast off his Jewish clothes, adopted a Turkish turban, and converted to Islam. Tens of thousands of his devout followers left the cult in utter disillusionment.

The prophecies of seers still resonate even today, influencing the lives of tens of millions of people worldwide. In the United States, William Miller declared that Doomsday would arrive on April 3, 1843. As news of his prophecy spread thoughout the United States, a spectacular meteor shower by chance lit up the night sky in 1833, one of the largest of its kind, further enhancing the influence of Miller's prophecy.

Tens of thousands of devout followers, called Millerites, awaited the coming of Armageddon. When 1843 came and went without the arrival of the End of Days, the Millerite movement split into several large groups. Because of the huge following amassed by the Millerites, each of these splinter groups would have a major impact on religion even today. One large piece of the Millerite movement regrouped in 1863 and changed their name to the Seventh-Day Adventist Church, which today has about 14 million baptized members. Central to their belief is the imminent Second Coming of Christ.

Another splinter group of Millerites later drifted toward the work of Charles Taze Russell, who pushed back the date of Doomsday to 1874. When that date also passed, he revised his prediction, based on analyses of the Great Pyramids of Egypt, this time to 1914. This group would later be called Jehovah's Witnesses, with a membership of over 6 million.

Other segments of the Millerite movement, however, continued to make predictions, hence precipitating further splits each time a prediction failed. One small splinter group of Millerites was called the Branch Davidians; they broke off from the Seventh-Day Adventists in the 1930s. They had a small commune in Waco, Texas, which fell under the charismatic influence of a young preacher named David Koresh, who spoke hypnotically of the end of the world. That group met a fiery end in their tragic encounter with the FBI in 1993, when a raging inferno consumed the compound, incinerating 76 members, including 27 children, and also Koresh.

Can We See the Future?

Can rigorous scientific tests prove that some individuals can see the future? In Chapter 12 we saw that time travel might be consistent with the laws of physics, but for an advanced, Type III civilization. But is precognition possible on Earth today?

Elaborate tests conducted at the Rhine Center seem to suggest that some people can see the future; that is, they can identify cards before they are unveiled. But repeated experiments have shown that the effect is very small, and often disappears when others try to duplicate the results.

In fact, precognition is difficult to reconcile with modern physics, because it violates causality, the law of cause and effect. Effects occur after the cause, not vice versa. All the laws of physics that have been found so far have causality built into them. A violation of causality would signal a major collapse of the foundations of physics. Newtonian mechanics is firmly based on causality. Newton's laws are so all-embracing that if you know the location and position of all the molecules in the universe, you can calculate the future motion of these atoms. Thus the future is calculable. In principle, Newtonian mechanics states that if you had a large enough computer, you could compute all future events. According to Newton, the universe is like a gigantic clock, wound up by God at the beginning of time, and ticking ever since according to His laws. There is no room for precognition in Newton's theory.

Backward in Time

When we discuss Maxwell's theory, however, the scenario becomes much more complicated. When we solve Maxwell's equations for light, we find not one but two solutions: a "retarded" wave, which represents the standard motion of light from one point to another; but also an "advanced" wave, where the light beam goes backward in time. This advanced solution comes from the future and arrives in the past!

For a hundred years when engineers have encountered this "advanced" solution that goes backward in time they have simply dismissed it as a mathematical curiosity. Since the retarded waves so accurately predicted the behavior of radio, microwaves, TV, radar, and X-rays, they simply threw the advanced solution out the window. The retarded waves were so spectacularly beautiful and successful that engineers simply ignored the ugly twin. Why tamper with success?

But for physicists, the advanced wave has been a nagging problem for the past century. Since Maxwell's equations are among the pillars of the modern age, any solution of these equations has to be taken very seriously, even if it entails accepting waves from the future. It seemed that it was impossible to totally ignore the advanced waves from the future. Why would nature, at this most fundamental level, give us such a bizarre solution? Was this a cruel joke, or was there a more profound meaning?

Mystics began to take an interest in these advanced waves, speculating that they would appear as messages from the future. Perhaps if we could somehow harness these waves, we might be able to send messages back to the past, and hence alert previous generations of events to come. We could, for example, send a message back to our grandparents in the year 1929, warning them to sell all their stocks before the Great Crash. Such advanced waves would not allow us personally to visit the past, as in time travel, but they would enable us to send letters and messages into the past to alert people of key events that would not yet have occurred.

These advanced waves were a mystery until they were studied by Richard Feynman, who was intrigued by the idea of going backward in time. After working on the Manhattan Project, which built the first atomic bomb, Feynman left Los Alamos and went to Princeton University to work under John Wheeler. Analyzing Dirac's original work on the electron, Feynman found something very strange. If he simply reversed the direction of time in Dirac's equation, the equation remained the same if he also reversed the electron charge. In other words, an electron going backward in time was the same as an antielectron going forward in time! Normally, a mature physicist might dismiss this interpretation, calling it just a trick, a mathematical sleight-of-hand with no meaning. Going backward in time did not seem to make any sense, yet Dirac's equations were clear on this point. In other words, Feynman had found the reason that nature allowed these backward-in-time solutions: they represented the motion of antimatter. If he had been an older physicist, Feynman might have thrown this solution out the window. But being a lowly graduate student, he decided to pursue his curiosity further.

As he continued to delve into this conundrum the young Feynman noticed something even stranger. Normally if an electron and an anti-electron collide, they annihilate one another and create a gamma ray. He drew this on a sheet of paper: two objects bumping into each other, turning into a burst of energy.

But then if you reversed the charge of the antielectron, it became an ordinary electron going backward in time. You could then rewrite the same diagram with the arrow of time reversed. It now appeared as if the electron went forward in time, then suddenly decided to reverse direction. The electron did a U-turn in time and was now going backward in time, releasing a burst of energy in the process. In other words, it's the *same* electron. The electron-antielectron annihilation process was just the same electron deciding to go backward in time!

So Feynman revealed the true secret of antimatter: *it's just ordinary matter going backward in time.* This simple observation immediately explained the puzzle that all particles have antiparticle partners: it's because all particles can travel backward in time, and hence masquerade as antimatter. (This interpretation is equivalent to the "Dirac sea," mentioned earlier, but it is simpler, and it is the explanation currently accepted today.)

Now let's say we have a lump of antimatter and it collides with ordinary matter, creating a huge explosion. There are now trillions of electrons and trillions of antielectrons being annihilated. But if we reversed the direction of the arrow for the antielectron, turning it into an electron going backward in time, this would mean that the same electron went zigzagging backward and forward trillions of times.

There was a further curious result: there must be just one electron in the lump of matter. The same electron went whizzing back and forth, zigzagging in time. Each time it did a U-turn in time it became antimatter. But if it did another U-turn in time then it turned into another electron.

(With his thesis adviser, John Wheeler, Feynman then speculated that perhaps the entire universe consisted of just one electron, zigzagging back and forth in time. Imagine that out of the chaos of the original big bang only a single electron was created. Trillions of years later, this single electron would eventually encounter the cataclysm of Doomsday, where it would make a U-turn and go backward in time, releasing a gamma ray in the process. Then it would go back to the original big bang, and then perform another U-turn. The electron would then make repeated zigzag journeys back and forth, from the big bang to Doomsday. Our universe in the twenty-first century is just a time slice of this electron's journey, in which we see trillions of electrons and antielectrons, that is, the visible universe. As strange as this theory may appear, it would explain a curious fact from the quantum theory: why all electrons are the same. In physics you cannot label electrons. There are no green electrons or Johnny electrons. Electrons have no individuality. You cannot "tag" an electron, like scientists sometimes tag animals in the wild to study them. Maybe the reason is that the entire universe consists of the same electron, just bouncing back and forth in time.)

But if antimatter is ordinary matter going back in time, then is it possible to send a message into the past? Is it possible to send today's *Wall Street Journal* back to yourself in the past, so you can make a killing on the stock market?

The answer is no.

If we treat antimatter as just another exotic form of matter and then perform an experiment with antimatter, there are no violations of causality. Cause and effect remain the same. If we now reverse the arrow of time for the antielectron, sending it backward in time, then we have only performed a mathematical operation. The physics remains the same. Nothing has changed physically. All experimental results remain the same. So it is absolutely valid to view the electron as going backward and forward in time. But each time the electron goes backward in time, it simply fulfills the past. So it appears as if the advanced solutions from the future are indeed necessary to have a consistent quantum theory, but they ultimately do not violate causality. (In fact, without these bizarre advanced waves, causality would be violated in the quantum theory. Feynman showed that if we add the contribution of the advanced and retarded waves, we find that the terms that might violate causality cancel precisely. Thus antimatter is essential to preserving causality. Without antimatter, causality might collapse.)

Feynman continued to pursue the germ of this crazy idea until it eventually blossomed into a complete quantum theory of the electron. His creation, quantum electrodynamics (QED), has been experimentally verified to one part in 10 billion, making it one of the most accurate theories of all time. It won him and his colleagues Julian Schwinger and Sin-Itiro Tomonaga the Nobel Prize in 1965.

(In Feynman's Nobel Prize acceptance speech, he said that as a youth he impulsively fell in love with these advanced waves from the future, like falling in love with a beautiful girl. Today that beautiful girl has matured into a grown woman and is the mother of many children. One of those children is his theory of quantum electrodynamics.)

Tachyons from the Future

In addition to advanced waves from the future (which have proven their utility over and over again in the quantum theory) there is yet another bizarre concept from the quantum theory that seems just as crazy, but perhaps not as useful. This is the idea of "tachyons," which appear regularly on *Star Trek*. Anytime the writers of *Star Trek* need some kind of new energy to perform some magical operation, they invoke tachyons.

Tachyons live in a strange world where everything travels faster than light. As tachyons lose energy, they travel faster, which violates common sense. In fact, if they lose all energy, they travel at infinite velocity. As tachyons gain energy, however, they slow down until they reach the speed of light.

What makes tachyons so strange is that they come with imaginary mass. (By "imaginary," we mean that their mass has been multiplied by the square root of minus one, or "i.") If we simply take Einstein's famous equations and replace "m" with "im," then something marvelous happens. All of a sudden particles travel faster than light.

This result gives rise to strange situations. If a tachyon travels through matter, it loses energy because it collides with atoms. But as it loses energy, it speeds up, which further increases its collisions with atoms. These collisions should cause it to lose more energy and hence accelerate even faster. As this creates a vicious cycle, the tachyon naturally attains infinite velocity all by itself!

(Tachyons are different from antimatter and negative matter. Antimatter has positive energy, travels at less than the speed of light, and can be created in our particle accelerators. It falls down under gravity, according to theory. Antimatter corresponds to ordinary matter going backward in time. Negative matter has negative energy and also travels less than the speed of light, but falls up under gravity. Negative matter has never been found in the laboratory. In large quantities, it can in theory be used to fuel time machines. Tachyons travel faster than light and have imaginary mass; it's not clear if they fall up or down under gravity. They, too, have not been found in the laboratory.)

As bizarre as tachyons are, they have been seriously studied by physicists, including the late Gerald Feinberg of Columbia University and George Sudarshan of the University of Texas at Austin. The problem is that no one has ever seen a tachyon in the laboratory. The key experimental evidence for tachyons would be a violation of causality. Feinberg even suggested that physicists examine a laser beam before it was switched on. If tachyons exist, then perhaps light from the laser beam could be detected even before the apparatus was turned on.

In science fiction stories tachyons are regularly used to send messages back to the past to seers. But if one examines the physics it is not clear if this is possible. Feinberg, for example, believed that the emission of a tachyon going forward in time was identical to the absorption of a negative-energy tachyon going backward in time (similar to the situation with regard to antimatter) and hence there was no violation of causality.

Science fiction aside, today the modern interpretation of tachyons is that they might have existed at the instant of the big bang, violating causality, but they don't exist anymore. In fact, they might have played an essential role in getting the universe to "bang" in the first place. In that sense, tachyons are essential for some theories of the big bang.

Tachyons have a peculiar property. When you put them into any theory, they destabilize the "vacuum," that is, the lowest energy state of a system. If a system has tachyons, it is in a "false vacuum," so the system is unstable and will decay down to the true vacuum.

Think of a dam that holds back the water in a lake. This represents the "false vacuum." Although the dam appears perfectly stable, there is an energy state that is lower than the dam. If a crack develops in the dam and the water comes bursting out of the dam break, the system attains the true vacuum as the water flows toward sea level.

In the same way, the universe before the big bang, it is believed, originally started off in the false vacuum, in which there were tachyons. But the presence of tachyons meant that this was not the lowest energy state, and hence the system was unstable. A tiny "rip" appeared in the fabric of space-time, representing the true vacuum. As the rip got larger, a bubble emerged. Outside the bubble the tachyons still exist, but inside the bubble the tachyons have all disappeared. As the bubble expands, we find the universe as we know it, without tachyons. This is the big bang.

One theory taken very seriously by cosmologists is that a tachyon, called the "inflation," started the original process of inflation. As we mentioned earlier, the inflationary universe theory states that the universe started off as a tiny bubble of space-time that underwent a turbocharged inflationary period. Physicists believe that the universe originally started off in the false vacuum state, where the inflation field was a tachyon. But the presence of a tachyon destabilized the vacuum, and tiny bubbles formed. Inside one of these bubbles the inflation field assumed the true vacuum state. This bubble then began to inflate rapidly, until it became our universe. Inside our bubble-universe the inflation has disappeared, so it can no longer be detected in our universe. So tachyons represent a bizarre quantum state in which objects go faster than light and perhaps even violate causality. But they disappeared a long time ago, and perhaps gave birth to the universe itself.

All this may sound like idle speculation that is not testable. But the theory of the false vacuum will get its first experimental test, starting in 2008, when the Large Hadron Collider is turned on outside Geneva, Switzerland. One of the key purposes of the LHC is to find the "Higgs boson," the last particle in the Standard Model, the one that has yet to be found. It is the last piece of this jigsaw puzzle. (The Higgs particle is so important but elusive that Nobel laureate Leon Lederman called it "The God Particle.")

The Higgs boson, physicists believe, originally started out as a tachyon. In the false vacuum, none of the subatomic particles had any mass. But its presence destabilized the vacuum, and the universe made a transition to a new vacuum, in which the Higgs boson turned into an ordinary particle. After the transition from a tachyon to an ordinary particle, the subatomic particles begin to have the masses that we measure in the laboratory today. Thus the discovery of the Higgs boson will not only complete the last missing piece of the Standard Model, it will also verify that the tachyon state once existed but has been transformed into an ordinary particle.

In summary, precognition is ruled out by Newtonian physics. The iron rule of cause and effect is never violated. In the quantum theory, new states of matter are possible, such as antimatter, which corresponds to matter going backward in time, but causality is not violated. In fact, in a quantum theory, antimatter is essential to restoring causality. Tachyons at first seem to violate causality, but physicists believe that their true purpose was to set off the big bang and hence they are not observable anymore.

Therefore precognition seems to be ruled out, at least for the foreseeable future, making it a Class III impossibility. It would set off a major shake-up in the very foundations of modern physics if precognition was ever proved in reproducible experiments.

Epilogue

THE FUTURE OF THE IMPOSSIBLE

There is nothing so big nor so crazy that one out of a million
technological societies may not feel itself driven to do,
provided it is physically possible.

–FREEMAN DYSON

Destiny is not a matter of chance–it is a matter of choice. It is not a
thing to be waited for–it is a thing to be achieved.

–WILLIAM JENNINGS BRYAN

Are there truths that will be forever beyond our grasp? Are there realms of knowledge that will be outside the capabilities of even an advanced civilization? Of all the technologies analyzed so far, only perpetual motion machines and precognition fall into the category of Class III impossibilities. Are there other technologies that are similarly impossible?

Pure mathematics abounds in theorems showing that certain things are truly impossible. One simple example is that it is impossible to trisect an angle using only a compass and ruler; this was proven back in 1837.

Even in simple systems such as arithmetic there are impossibilities. As I mentioned earlier, it is impossible to prove all the true statements in arithmetic within the postulates of arithmetic. Arithmetic is incomplete. There will always be true statements in arithmetic that can be proven only if one moves to a much larger system that includes arithmetic as a subset.

Although some things in mathematics are impossible, it is always dangerous to declare that something is absolutely impossible in the physical sciences. Let me remind you of a speech given by Nobel laureate Albert A. Michelson in 1894 at the dedication of the Ryerson Physical Lab at the University of Chicago, in which he declared that it was impossible to discover any new physics: "The more important fundamental laws and facts of physical science have all been discovered, and these are now so firmly established that the possibility of their ever being supplanted in consequence of new discoveries is exceedingly remote . . . Our future discoveries must be looked for in the sixth place of decimals."

His remarks were uttered on the eve of some of the greatest upheavals in scientific history, the quantum revolution of 1900, and the relativity revolution of 1905. The point is that things that are impossible today violate the known laws of physics, but the laws of physics, as we know them, can change.

In 1825 the great French philosopher Auguste Comte, writing in *Cours de Philosophie,* declared that it was impossible for science to determine what the stars were made of. This seemed like a safe bet at the time, since nothing was known about the nature of stars. They were so distant that it was impossible to visit them. Yet just a few years after he made this claim, physicists (using spectroscopy) declared that the sun was made of hydrogen. In fact, we now know that by analyzing the spectral lines from stars emitted billions of years ago it is possible to determine the chemical nature of most of the universe.

Comte challenged the world of science by making a list of other "impossibilities":

- He claimed that the "ultimate structure of bodies must always transcend our knowledge." In other words, it was impossible to know the true nature of matter.

- He thought that mathematics could never be used to explain biology and chemistry. It was impossible, he claimed, to reduce these sciences to mathematics.
- He thought that it was impossible that the study of heavenly bodies would have any impact on human affairs.

In the nineteenth century it was reasonable to propose these "impossibilities" since so little was known about fundamental science. Almost nothing was known about the secrets of matter and life. But today we have the atomic theory, which has opened up a whole new realm of scientific investigation into the structure of matter. We know about DNA and the quantum theory, which have unraveled the secrets of life and chemistry. We also know about meteor impacts from space, which have not only influenced the course of life on Earth, but have helped to shape its very existence.

Astronomer John Barrow notes, "Historians still debate the suggestion that Comte's views were partly responsible for the subsequent decline in French science."

Mathematician David Hilbert, in rejecting Comte's claims, wrote, "The true reason, according to my thinking, why Comte could not find an unsolvable problem lies in the fact that there is no such thing as an unsolvable problem."

But today some scientists are raising a new set of impossibilities: we will never know what happened before the big bang (or why it "banged" in the first place), and we will never achieve a "theory of everything."

Physicist John Wheeler commented on the first "impossible" question when he wrote: "Two hundred years ago, you could ask anybody, 'Can we someday understand how life came into being?' and he would have told you, 'Preposterous! Impossible!' I feel the same way about the question, 'Will we ever understand how the universe came into being?' "

Astronomer John Barrow adds, "The speed at which light travels is limited and so, therefore, is our knowledge of the structure of the Universe. We cannot know whether it is finite or infinite, whether it had a beginning or will have an end, whether the structure of physics is the same everywhere, or whether the Universe is ultimately a tidy or an untidy place . . . All the great questions about the nature of the Universe–from its beginning to its end–turn out to be unanswerable."

Barrow is correct in saying that we will never know, with absolute certainty, the true nature of the universe, in all its glory. But it is possible to incrementally chip away at these eternal questions and come tantalizingly close. Instead of representing the absolute boundaries of our knowledge, these "impossibilities" may perhaps better be seen as the challenges awaiting the next generation of scientists. These limits are like piecrusts, made to be broken.

Detecting the Pre–Big Bang Era

In the case of the big bang, a new generation of detectors is being built that could settle some of these eternal questions. Today our radiation detectors in outer space can only measure the microwave radiation emitted 300,000 years after the big bang, when the first atoms formed. It is impossible to use this microwave radiation to probe earlier than 300,000 years after the big bang, since radiation from the original fireball was too hot and random to yield useful information.

But if we analyze other types of radiation we may be able to get even closer to the big bang. Tracking neutrinos, for example, can take us closer to the instant of the big bang (neutrinos are so elusive that they can travel through an entire solar system made of solid lead). Neutrino radiation could take us within a few seconds after the big bang.

But perhaps the ultimate secret of the big bang will be revealed by examining "gravity waves," waves that move along the fabric of space-time. As physicist Rocky Kolb of the University of Chicago says, "By measuring the properties of the neutrino background we can look back to one second after the Bang. But gravitational waves from [the] inflation area are relics of the universe 10^{-35} seconds after the bang."

Gravity waves were first predicted by Einstein in 1916; they may eventually become the most important probe for astronomy. Historically each time a new form of radiation was harnessed, a new era in astronomy was opened up. The first form of radiation was visible light, used by Galileo to investigate the solar system. The second form of radiation was radio waves, which eventually enabled us to probe the centers of galaxies to find black holes. Gravity wave detectors may unveil the very secrets of creation.

In some sense gravity waves have to exist. To see this, consider the age-old question: what happens if the sun suddenly disappears? According to Newton, we would feel the effects immediately. The Earth would be instantly thrown out of its orbit and plunged into darkness. This is because Newton's law of gravity does not take into account velocity, and hence forces act instantly throughout the universe. But according to Einstein, nothing can travel faster than light, so it would take eight minutes for the information about the sun's disappearance to reach the Earth. In other words, a spherical "shock wave" of gravity would emerge from the sun and eventually hit the Earth. Outside this sphere of gravity waves, it would appear as if the sun were still shining normally, because information about the disappearance of the sun would not have reached Earth. Inside this sphere of gravity waves, however, the sun would have already disappeared, as the expanding shock wave of gravity waves travels at the speed of light.

Another way to see why gravity waves must exist is to visualize a large bed sheet. According to Einstein, space-time is a fabric that can be warped or stretched, like a curved bed sheet. If we grab a bed sheet and shake it rapidly we see that waves ripple along the surface of the bed sheet and travel at a definite velocity. In the same way, gravity waves can be viewed as waves traveling along the fabric of space-time.

Gravity waves are among the fastest-moving topics in physics today. In 2003 the first large-scale gravity wave detectors became operational—called LIGO (Laser Interferometer Gravitational Wave Observatory), measuring 2.5 miles in length, one is based in Hanford, Washington, and another in Livingston Parish, Louisiana. It is hoped that LIGO, at a cost of $365 million, will be able to detect radiation from colliding neutron stars and black holes.

The next big leap will take place in 2015, when an entirely new generation of satellites will be launched that will analyze gravitational radiation in outer space from the instant of creation. The three satellites that make up LISA (Laser Interferometer Space Antenna), a joint project of NASA and the European Space Agency, will be sent into orbit around the sun. These satellites will be capable of detecting gravitational waves emitted less than a trillionth of a second after the big bang. If a gravity wave from the big bang still circulating around the universe hits one of the satellites, it will disturb the laser beams, and this disturbance can then be measured in a precise way, giving us "baby pictures" of the instant of creation itself.

LISA consists of three satellites circling the sun arranged in a triangle, each connected by laser beams 3 million miles long, making it the largest instrument of science ever created. This system of three satellites will orbit the sun about 30 million miles from the Earth.

Each satellite will emit a laser beam with only half a watt of power. By comparing the laser beams coming from the other two satellites, each satellite will be able to construct an interference pattern of light. If a gravity wave disturbs the laser beams, it will change the interference pattern, and the satellite will be able to detect this disturbance. (The gravity wave does not make the satellites vibrate. It actually creates a distortion in the space between the three satellites.)

Although the laser beams are very weak, their accuracy will be astounding. They will be able to detect vibrations to within one part in a billion trillion, corresponding to a shift 1/100 the size of an atom. Each laser beam will be able to detect a gravity wave from a distance of 9 billion light-years, which covers most of the visible universe.

LISA has the sensitivity to potentially differentiate between several "pre–big bang" scenarios. One of the hottest topics in theoretical physics today is calculating the characteristics of the pre–big bang universe. At present, inflation can describe quite well how the universe evolved once the big bang took place. But inflation cannot explain why the big bang took place in the first place. The goal is to use these speculative models of the pre–big bang era to calculate the gravity radiation emitted by the big bang. Each of the various pre–big bang theories makes different predictions. The big bang radiation predicted by the

Big Splat theory, for example, differs from the radiation predicted by some of the inflation theories, so LISA might be able to rule out several of these theories. Obviously, these pre–big bang models cannot be tested directly, since they involve understanding the universe before the creation of time itself, but we can test them indirectly since each of these theories predicts a different radiation spectrum emerging afterward from the big bang.

Physicist Kip Thorne writes, "Sometime between 2008 and 2030, gravitational waves from the Big Bang singularity will be discovered. There will ensue an era, lasting at least until 2050 . . . These efforts will reveal intimate details of the Big Bang singularity, and will thereby verify that some version of string theory is the correct quantum theory of gravity."

If LISA is unable to differentiate between different pre–big bang theories, its successor, the Big Bang Observer (BBO) might. It is tentatively scheduled for launch in 2025. The BBO will be able to scan the entire universe for all binary systems involving neutron stars and black holes with mass less than one thousand times the mass of the sun. But its main goal is to analyze gravity waves emitted during the inflationary phase of the big bang. In this sense, the BBO is specifically designed to probe the predictions of the inflationary big bang theory.

The BBO is somewhat similar to LISA in design. It will consist of three satellites moving together in an orbit around the sun, separated from each other by 50,000 kilometers (these satellites will be much closer to one another than LISA's satellites). Each satellite will be able to fire a 300-watt laser beam. BBO will be able to probe gravity wave frequencies between LIGO and LISA, filling an important gap. (LISA can detect gravity waves from 10 to 3,000 hertz, while LIGO can detect gravity waves of frequency 10 microhertz to 10 millihertz. BBO will be able to detect frequencies that include both ranges.)

"By 2040 we will have used those laws [of quantum gravity] to produce high-confidence answers to many deep and puzzling questions," Thorne writes, "including . . . What came before the Big Bang singularity, or was there even such a thing as a 'before'? Are there other universes? And if so, how are they related to or connected to our own universe? . . . Do the laws of physics permit highly advanced civilizations to create and maintain wormholes for interstellar travel, and to create time machines for backward time travel?"

The point is that in the next few decades there should be enough data pouring in from gravity wave detectors in space to differentiate between the various pre–big bang theories.

THE END OF THE UNIVERSE

The poet T. S. Eliot asked the question, Will the universe die with a bang or a whimper? Robert Frost asked, Will we all perish in fire or ice? The latest evidence points to the universe dying in a Big Freeze, in which temperatures will reach near absolute zero and all intelligent life will be extinguished. But can we be sure?

Some have raised another "impossible" question. How will we ever know the ultimate fate of the universe, they ask, since this event is trillions upon trillions of years in the future? Scientists believe that "dark energy" or the energy of the vacuum seems to be pushing the galaxies apart at an ever increasing rate, indicating that the universe seems to be in a runaway mode. Such an expansion would cool the temperature of the universe and ultimately lead to the Big Freeze. But is this expansion temporary? Could it reverse itself in the future?

For example, in the Big Splat scenario, in which two membranes collide and create the universe, it appears as if the membranes can collide periodically. If so, then the expansion that appears to lead to a Big Freeze is only a temporary state that will reverse itself.

What is driving the current acceleration of the universes is dark energy, which in turn is probably caused by the "cosmological constant." The key, therefore, is to understand this mysterious constant, or the energy of the vacuum. Does the constant vary with time, or is it really a constant? At present, no one knows for sure. We know from the WMAP satellite currently orbiting the Earth that this cosmological constant seems to be driving the current acceleration of the universe, but we don't know if it is permanent or not.

This problem is actually an old one, dating back to 1916 when Einstein first introduced the cosmological constant. Soon after proposing general relativity the previous year, he worked out the cosmological implications of his own theory. Much to his surprise, he found that the universe was dynamic, that it either expanded or contracted. But this idea seemed to contradict the data.

Einstein was encountering the Bentley paradox, which had bedeviled even Newton. Back in 1692 the Reverend Richard Bentley wrote Newton an innocent letter with a devastating question. If Newton's gravity was always attractive, Bentley asked, then why doesn't the universe collapse? If the universe consists of a finite collection of stars that mutually attract each other, then the stars should come together and the universe should collapse into a fireball! Newton was deeply troubled by this letter, since it pointed out a key flaw in his theory of gravity: *any theory of gravity that is attractive is inherently unstable.* Any finite collection of stars will inevitably collapse under gravity.

Newton wrote back that the only way to create a stable universe was to have an infinite and uniform collection of stars, with each star being pulled in all directions, so that all the forces cancel out. It was a clever solution, but Newton was smart enough to realize that such stability was deceptive. Like a house of cards, the tiniest of vibrations would cause the whole thing to collapse. It was "metastable"; that is, it was temporarily stable until the slightest perturbations caused it to crash. Newton concluded that God was necessary to periodically nudge the stars a bit so the universe did not collapse.

In other words, Newton saw the universe as a gigantic clock, wound up by God at the beginning of time and obeying Newton's laws. It has been ticking automatically ever since, without divine intervention. However, according to Newton, God was necessary to tweak the stars once in a while so the universe did not collapse into a fireball.

When Einstein stumbled on the Bentley paradox in 1916, his equations correctly told him that the universe was dynamic–either expanding or contracting–and that a static universe was unstable and would collapse due to gravity. But the astronomers insisted at that time that the universe was static and unchanging. So Einstein, bowing to the observations of the astronomers, added the cosmological constant, an antigravity force that pushed the stars apart to balance the gravitational pull causing the universe to collapse. (This antigravity force corresponded to the energy contained within the vacuum. In this picture even the vast emptiness of space contains large quantities of invisible energy.) This constant would have to be chosen very precisely in order to cancel out the attractive force of gravity.

Later, when Edwin Hubble showed in 1929 that the universe was, in fact, expanding, Einstein would say that the cosmological constant was his "greatest blunder." Yet now, seventy years later, it appears as if Einstein's "blunder," the cosmological constant, could in fact be the largest source of energy in the universe, making up 73 percent of the matter-energy content of the universe. (By contrast, the higher elements that make up our bodies constitute only .03 percent of the universe.) Einstein's blunder will likely determine the ultimate fate of the universe.

But where does this cosmological constant come from? At present no one knows. At the beginning of time, the antigravity force was perhaps large enough to cause the universe to inflate, creating the big bang. Then it suddenly disappeared, for reasons that are unknown. (The universe was still expanding during this period, but at a slower pace.) And then, about eight billion years after the big bang, the antigravity force resurfaced again, causing the galaxies to push out and causing the universe to accelerate once again.

So is it "impossible" to determine the ultimate fate of the universe? Perhaps not. Most physicists believe that quantum effects ultimately determine the size of the cosmological constant. A naïve calculation, using a primitive version of the quantum theory, shows that the cosmological constant is off by a factor of 10^{120}. This is the greatest mismatch in the history of science.

But there is also a consensus among physicists that this anomaly simply means that we need a theory of quantum gravity. Since the cosmological constant arises via quantum corrections, it is necessary to have a theory of everything–a theory that will allow us to calculate not just the Standard Model, but also the value of the cosmological constant, which will determine the ultimate fate of the universe.

So a theory of everything is necessary to determine the ultimate fate of the universe. The irony is that some physicists believe that it is impossible to attain a theory of everything.

A Theory of Everything?

As I mentioned earlier, string theory is the leading candidate for a "theory of everything," yet there are opposing camps on whether the string theory lives up to this claim. On the one hand, people like MIT professor Max Tegmark write, "In 2056, I think you'll be able to buy a T-shirt on which are printed equations describing the unified physical laws of our universe." On the other hand, there is an emerging band of determined critics who claim that the string bandwagon has yet to deliver. No matter how many breathless articles or TV documentaries are produced concerning string theory, it has yet to produce a single testable fact, some say. It's a theory of nothing, rather than a theory of everything, claim the critics. The debate heated up considerably in 2002 when Stephen Hawking switched sides, quoting the incompleteness theorem, and said that a theory of everything might even be mathematically impossible.

It's not surprising that the debate has pitted physicist against physicist, because the goal is so lofty, if elusive. The quest to unify all the laws of nature has tantalized and lured philosophers and physicists alike for millennia. Socrates himself once said, "It seemed to me a superlative thing–to know the explanation of everything, why it comes to be, why it perishes, why it is."

The first serious proposal for a theory of everything dates back to about 500 BC, when the Greek Pythagoreans are credited with deciphering the mathematical laws of music. By analyzing the nodes and vibrations of a lyre string, they showed that music obeyed remarkably simple mathematics. They then speculated that all of nature could be explained in the harmonies of the lyre string. (In some sense, string theory brings back the dream of the Pythagoreans.)

In modern times nearly all the giants of twentieth-century physics tried their luck with a unified field theory. But, as Freeman Dyson cautions, "The ground of physics is littered with the corpses of unified theories."

In 1928 the *New York Times* ran the sensational headline "Einstein on verge of great discovery; resents intrusion." The news story helped spark a media feeding frenzy over a theory of everything that was whipped up to a feverish pitch. Headlines blared "Einstein is amazed at stir over theory. Holds 100 journalists at bay for a week." Scores of journalists swarmed around his home in Berlin, maintaining a nonstop vigil, waiting to catch a glimpse of the genius and grab a headline. Einstein was forced to go into hiding.

Astronomer Arthur Eddington wrote to Einstein: "You may be amused to hear that one of our great department stores in London (Selfridges) has posted on its window your paper (the six pages pasted up side by side) so that passers-by can read it all through. Large crowds gather around to read it." (In 1923 Eddington proposed his own unified field theory on which he worked tirelessly for the rest of his life, until he died in 1944.)

In 1946 Erwin Schrödinger, one of the founders of quantum mechanics, held a press conference to propose his unified field theory. Even Ireland's Prime Minister, Eamon De Valera, showed up. When a reporter asked him what he would do if his theory was wrong, Schrödinger replied, "I believe I am right. I shall look like an awful fool if I am wrong." (Schrödinger was humiliated when Einstein politely pointed out the errors in his theory.)

The harshest of all critics of unification was physicist Wolfgang Pauli. He chided Einstein, saying, "What God has torn asunder, let no man put together." He mercilessly put down any half-baked theory with the quip: "It's not even wrong." So it is ironic that the supreme cynic Pauli himself caught the bug. In the 1950s he proposed his own unified field theory with Werner Heisenberg.

In 1958 Pauli presented the Heisenberg-Pauli unified theory at Columbia University. Niels Bohr was in the audience, and he was not impressed. Bohr stood up and said, "We in the back are convinced that your theory is crazy. But what divides us is whether your theory is crazy enough." The criticism was crushing. Since all the obvious theories had been considered and rejected, the true unified field theory must be a dazzling departure from the past. The Heisenberg-Pauli theory was simply too conventional, too ordinary, too sane to be the true theory. (That year Pauli was disturbed when Heisenberg commented on a radio broadcast that only a few technical details were left in their theory. Pauli sent his friends a letter with a blank rectangle, with the caption, "This is to show the world I can paint like Titian. Only technical details are missing.")

Criticisms of String Theory

Today the leading (and only) candidate for a theory of everything is string theory. But, again, a backlash has arisen. Opponents claim that to get a tenured position at a top university you have to work on string theory. If you don't you will be unemployed. It's the fad of the moment, and it's not good for physics.

I smile when I hear this criticism, because physics, like all human endeavors, is subject to fads and fashions. The fortunes of great theories, especially on the cutting edge of human knowledge, can rise and fall like hemlines. In fact, years ago the tables were turned; string theory was historically an outcast, a renegade theory, the victim of the bandwagon effect.

String theory was born in 1968, when two young postdocs, Gabriel Veneziano and Mahiko Suzuki, stumbled on a formula that seemed to describe the collisions of subatomic particles. Quickly it was discovered that this marvelous formula could be derived by the collision of vibrating strings. But by 1974 the theory was dead in its tracks. A new theory, quantum chromodynamics (QCD), or the theory of quarks and the strong interaction, was a juggernaut flattening all other theories.

People left string theory in droves to work on QCD. All the funding, jobs, and recognition went to physicists working on the quark model.

I remember those dark years well. Only the foolhardy or the stubborn persisted in working on string theory. And when it became known that these strings could vibrate only in ten dimensions, the theory became the butt of jokes. String pioneer John Schwarz at Cal Tech would sometimes bump into Richard Feynman in the elevator. Ever the joker, Feynman would ask, "Well, John, and how many dimensions are you in today?" We used to joke that the only place to find a string theorist was in the unemployment line. (Nobel laureate Murray Gell-Mann, founder of the quark model, once confided to me that he took pity on string theorists and created a "nature preserve for endangered string theorists" at Cal Tech so people like John wouldn't lose their jobs.)

Given that today so many young physicists are rushing to work on string theory, Steve Weinberg has written, "String theory provides our only present source of candidates for a final theory–how could anyone expect that many of the brightest young theorists would *not* work on it?"

Is String Theory Untestable?

One major criticism of string theory today is that it is untestable. It would take an atom smasher the size of the galaxy to test this theory, critics claim.

But this criticism neglects the fact that most science is done indirectly, not directly. No one has ever visited the sun to do a direct test, but we know it is made of hydrogen because we can analyze its spectral lines.

Or take black holes. The theory of black holes dates back to 1783, when John Michell published an article in the *Philosophical Transactions of the Royal Society*. He claimed that a star could be so massive that "all light emitted from such a body would be made to return to it by its own proper gravity." Michell's "dark star" theory languished for centuries because a direct test was impossible. In 1939 Einstein even wrote a paper showing that such a dark star could not form by natural means. The criticism was that these dark stars were inherently untestable because they were, by definition, invisible. Yet today the Hubble Space Telescope has given us gorgeous evidence of black holes. We now believe that billions of them could lurk in the hearts of galaxies; scores of wandering black holes could exist in our own galaxy. But the point is that the evidence for black holes is all indirect; that is, we have gathered information about black holes by analyzing the accretion disk that swirls around them.

Furthermore, many "untestable" theories ultimately become testable. It took two thousand years to prove the existence of atoms after they were first proposed by Democritus. Nineteenth-century physicists such as Ludwig Boltzmann were hounded to death for believing in that theory, yet today we have gorgeous photographs of atoms. Pauli himself introduced the concept of the neutrino in 1930, a particle so elusive it can pass through blocks of solid lead the size of an entire star system and not be absorbed. Pauli said, "I have committed the ultimate sin; I have introduced a particle that can never be observed." It was "impossible" to detect the neutrino, so it was considered little more than science fiction for several decades. Yet today we can produce beams of neutrinos.

There are, in fact, a number of experiments that will provide, physicists hope, the first indirect tests of string theory:

- The Large Hadron Collider (LHC) might be powerful enough to produce "sparticles," or superparticles, which are the higher vibrations predicted by superstring theory (as well as by other supersymmetric theories).
- As I mentioned earlier, in 2015 the Laser Interferometer Space Antenna (LISA) will be launched in space. LISA and its successor, the Big Bang Observer, may be sensitive enough to test several "pre–big bang" theories, including versions of the string theory.
- A number of labs are investigating the presence of

higher dimensions by looking at deviations from Newton's famed inverse-square law at the millimeter scale. (If there is a fourth spatial dimension, then gravity should fall by the inverse cube, not the inverse square.) The latest version of string theory (M-theory) predicts there are eleven dimensions.

• Many labs are trying to detect dark matter, since the Earth is moving in a cosmic wind of dark matter. String theory makes specific, testable predictions about the physical properties of dark matter because dark matter is probably a higher vibration of the string (e.g., the photino).

• It is hoped that a series of additional experiments (e.g., on neutrino polarization in the south pole) will detect the presence of mini–black holes and other strange objects by analyzing anomalies in cosmic rays, whose energies can easily exceed those of the LHC. Cosmic ray experiments and the LHC will open a new, exciting frontier beyond the Standard Model.

• And there are some physicists who hold out the possibility that the big bang was so explosive that perhaps a tiny superstring was blown up into astronomical proportions. As physicist Alexander Vilenkin of Tufts University writes, "A very exciting possibility is that superstrings . . . can have astronomical dimensions . . . We would then be able to observe them in the sky and directly test superstring theory." (The probability of finding a huge, relic superstring that was blown up during the big bang is quite small.)

Is Physics Incomplete?

In 1980 Stephen Hawking helped to spark interest in a theory of everything with his lecture entitled "Is the End in Sight for Theoretical Physics?" in which he said, "We may see a complete theory within the lifetime of some of those present here." He claimed that there was a fifty-fifty chance that the final theory would be found in the next twenty years. But when the year 2000 arrived and there was no consensus on the theory of everything, he changed his mind and said that there was a fifty-fifty chance of finding it in another twenty years.

Then in 2002 Hawking changed his mind once again, declaring that Gödel's incompleteness theorem may suggest a fatal flaw in his original line of thinking. He wrote, "Some people will be very disappointed if there is not an ultimate theory that can be formulated as a finite number of principles. I used to belong to that camp, but I have changed my mind . . . Gödel's theorem ensured there would always be a job for mathematicians. I think M-theory will do the same for physicists."

His argument is an old one: since mathematics is incomplete and the language of physics is mathematics, there will always be true physical statements that are forever beyond our reach, and hence a theory of everything is not possible. Since the incompleteness theorem killed off the Greek dream of proving all true statements in mathematics, it will also put a theory of everything forever beyond our reach.

Freeman Dyson said it eloquently when he wrote, "Gödel proved the world of pure mathematics is inexhaustible; no finite set of axioms and rules of inference can ever encompass the whole of mathematics . . . I hope that an analogous situation exists in the physical world. If my view of the future is correct, it means that the world of physics and astronomy is also inexhaustible; no matter how far we go into the future, there will always be new things happening, new information coming in, new worlds to explore, a constantly expanding domain of life, consciousness, and memory."

Astrophysicist John Barrow summarizes this logic this way: "Science is based on mathematics; mathematics cannot discover all truths; therefore science cannot discover all truths."

Such an argument may or may not be true, but there are potential flaws. Professional mathematicians for the most part ignore the incompleteness theorem in their work. This is because the incompleteness theorem begins by analyzing statements that refer to themselves; that is, they are self-referential. For example, statements like the following are paradoxical:

> This sentence is false.
> I am a liar.
> This statement cannot be proven.

In the first case, if the sentence is true, it means it is false. If the sentence is false, then the statement is true. Likewise, if I am telling the truth, then I am telling a lie; and if I am telling a lie, then I am telling the truth. In the last case, if the sentence is true, then it cannot be proven to be true.

(The second statement is the famous liar's paradox. The Cretan philosopher Epimenides used to illustrate this paradox by saying, "All Cretans are liars." However, Saint Paul missed the point entirely and wrote, in his epistle to Titus, "One of Crete's own prophets has said it, 'Cretans are always liars, evil brutes, lazy gluttons.' He has surely told the truth.")

The incompleteness theorem builds on statements such as "This sentence cannot be proven using the axioms of arithmetic" and creates a sophisticated web of these self-referential paradoxes.

Hawking, however, uses the incompleteness theorem to show that a theory of everything cannot exist. He claims that the key to Gödel's incompleteness theorem is that mathematics is self-referential, and physics suffers from this disease as well. Since the observer cannot be separated from the observation process, it means that physics will always refer to itself, since we cannot leave the universe. In the final analysis, the observer is also made of atoms and molecules, and hence must be an integral part of the experiment he is performing.

But there is a way to avoid Hawking's criticism. To avoid the paradoxes inherent in Gödel's theorem, professional mathematicians today simply state that their work excludes all self-referential statements. They can then circumvent the incompleteness theorem. To a large degree, the explosive development of mathematics since Gödel's time has been accomplished simply by ignoring the incompleteness theorem, that is, by postulating that recent work makes no self-referential statements.

In the same way it may be possible to construct a theory of everything that can explain every known experiment independent of the observer/observed dichotomy. If such a theory of everything can explain everything from the origin of the big bang to the visible universe that we see around us, then it becomes academic how we describe the interaction between the observer and observed. In fact, one criterion for a theory of everything should be that its conclusions are totally independent of how we make the split between the observer and the observed.

Furthermore, nature may be inexhaustible and limitless, even if it is based on a handful of principles. Consider a chess game. Ask an alien from another planet to figure out the rules of chess simply by watching the game. After a while the alien can figure out how pawns, bishops, and kings move. The rules of the game are finite and simple. But the number of possible games is truly astronomical. In the same way the rules of nature may also be finite and simple, but the applications of those rules may be inexhaustible. Our goal is to find the rules of physics.

In some sense we already have a complete theory of many phenomena. No one has ever seen a defect in Maxwell's equations for light. The Standard Model is often called a "theory of almost everything." Assume for the moment that we can shut off gravity. Then the Standard Model becomes a perfectly sound theory of all phenomena besides gravity. The theory may be ugly, but it works. Even in the presence of the incompleteness theorem, we have a perfectly reasonable theory of everything (besides gravity).

To me it is truly remarkable that on a single sheet of paper one can write down the laws that govern all known physical phenomena, covering forty-three orders of magnitude, from the farthest reaches of the cosmos over 10 billion light-years away to the microworld of quarks and neutrinos. On that sheet of paper would be just two equations, Einstein's theory of gravity and the Standard Model. To me this reveals the ultimate simplicity and harmony of nature at the fundamental level. The universe could have been perverse, random, or capricious. And yet it appears to us to be whole, coherent, and beautiful.

Nobel laureate Steve Weinberg compares our search for a theory of everything to the search for the North Pole. For centuries the ancient mariners worked with maps in which the North Pole was missing. All compass needles and charts pointed to this missing piece of the map, yet no one had actually visited it. In the same way, all our data and theories point to a theory of everything. It is the missing piece of our equations.

There will always be things that are beyond our grasp, that are impossible to explore (such as the precise position of an electron, or the world existing beyond the reach of the speed of light). But the fundamental laws, I believe, are knowable and finite. And the coming years in physics could be the most exciting of all, as we explore the universe with a new generation of particle accelerators, space-based gravity wave detectors, and other technologies. We are not at the end, but at the beginning of a new physics. But whatever we find, there will always be new horizons continually awaiting us.

NOTES

Preface

Page xv: This has happened several times... The reason that this is true is because of the quantum theory. When we add all possible quantum corrections to a theory (a tedious process called "renormalization") we find that phenomena that were previously forbidden, at the classical level, reenter the calculation. This means that unless something is explicitly forbidden (by a conservation law, for example) then it reenters into the theory when quantum corrections are added.

2: Invisibility

Page 17: Invisibility played a central part in Plato's theory... Plato wrote, "No man would keep his hands off what was not his own when he could safely take what he liked out of the market, or go into houses and lie with anyone at his pleasure, or kill or release from prison whom he would, and in all respects be like a God among men ... If you could imagine anyone obtaining this power of becoming invisible, and never doing any wrong or touching what was another's, he would be thought by the lookers-on to be the most wretched idiot..."

Page 21: Nathan Myhrvold, former chief technology officer at Microsoft... Nathan Myhrvold, *New Scientist Magazine,* November 18, 2006, p. 69.

Page 23: That's why he now declines... Josie Glausiusz, *Discover Magazine,* November 2006.

Page 25: "Such a lens would offer..." "Metamaterials found to work for visible light," Eurekalert, www.eurekalert.org/pub_releases/2007-01, 2007. Also, *New Scientist Magazine,* December 18, 2006.

3: Phasers and Death Stars

Page 36: During World War II, the Nazis... The Nazis also sent a team to India to investigate some ancient mythological claims of the Hindus (similar to the plot line in *Raiders of the Lost Ark*). The Nazis were interested in the writings of the Mahabharata, which described strange, powerful weapons, including flying craft.

Page 36: Weapons created from focused light beams... Movies like this have also spread a number of misconceptions about lasers. Laser beams are actually invisible unless they are scattered by particles in the air. So when Tom Cruise had to navigate through a maze of laser beams in *Mission Impossible,* the lattice of laser beams should have been invisible, not red. Also in many ray gun battles in the movies you can actually see the laser pulses zip across a room, which is impossible, since laser light travels at the speed of light, 186,000 miles per second.

Page 37: Writing about Einstein, Planck said, "That he may sometimes have missed the target..." Asimov and Schulman, p. 124.

4: Teleportation

Page 53: The earliest mention of teleportation can be found... The best recorded example of teleportation is dated October 24, 1593, when Gil Perez, a palace guard in the Philippine military guarding the governor in Manila, suddenly appeared in the Plaza Mayor of Mexico City. Dazed and confused, he was arrested by the Mexican authorities who thought he was in league with Satan. When he was brought before the Most Holy Tribunal of the Inquisition, all he could say in his defense was that he had disappeared from Manila to Mexico "in less time than it takes a cock to crow." (As incredible as the historic accounts of this incident may be, historian Mike Dash has noted that the earliest records of Perez's disappearance date from a century after his disappearance, and hence cannot be fully trusted.)

Page 54: Sir Arthur Conan Doyle, best known for his Sherlock Holmes novels... Doyle's early work was renowned for the methodical, logical thinking typical of the medical profession, as seen in the superb deductions of Sherlock Holmes. So why did Doyle decide to shift sharply away from the cold, rational logic of Mr. Holmes to the seat-of-your-pants, harrowing adventures of Professor Challenger, who delved into the forbidden worlds of mysticism, the occult, and the fringes of science? The author was profoundly changed by the sudden, unexpected deaths of several close relatives in World War I, including his beloved son Kingsley, his brother, two brothers-in-law, and two nephews. These losses would leave a deep, lasting emotional scar on him.

Depressed by their tragic deaths, Doyle embarked on a lifelong fascination with the world of the occult, believing perhaps that he might be able to communicate with the dead via spiritualism. He abruptly shifted from the world of rational, forensic science into mysticism, and went on to give famous lectures around the world about unexplained psychic phenomena.

Page 58: This uncertainty was finally codified by Heisenberg... More precisely, the Heisenberg uncertainty principle says that the uncertainty in the position of a particle, multiplied by the uncertainty in its momentum, must be greater than or equal to Planck's constant divided by 2π. Or the product of the uncertain in a particle's energy times the uncertainty in its time must also be greater than or equal to Planck's constant divided by 2 π. If we let Planck's constant go to zero, then this reduces to ordinary Newtonian theory, in which all uncertainties are zero.

The fact that you cannot know the position, momentum, energy, or time of an electron prompted Tryggvi Emilsson to wisecrack, "Historians have concluded that Heisenberg must have been contemplating his love life when he discovered the Uncertainty Principle:–When he had the time, he didn't have the energy and,–when the moment was right, he couldn't figure out the position.") Barrow, *Between Inner Space and Outer Space,* p. 187.

Page 58: "For my part, at least, I am convinced that He doesn't throw dice." Kaku, *Einstein's Cosmos,* p. 127.

Page 60: Bemoaning the undeniable experimental successes of the quantum theory, Einstein wrote, . . . Asimov and Schulman, p. 211.

Page 62: Everything changed in 1993, when scientists at IBM . . . Assume for the moment that macroscopic objects, including people, can be teleported. This raises subtle philosophical and theological questions about the existence of a "soul" if a person's body is teleported. If you are teleported to a new location, does your soul also move with you?

Some of these ethical questions were explored in James Patrick Kelley's novel *Think Like a Dinosaur.* In this tale a woman is teleported to another planet, but there is a problem with the transmission. Instead of the original body being destroyed, the original remains untouched, with all her emotions intact. Suddenly, there are two copies of her. Naturally, when the copy is told to enter the teleportation machine to be disintegrated she refuses. This creates a crisis, because the cold-blooded aliens, who provided the technology in the first place, view this as a purely practical matter to "balance the equation," while emotion-prone humans are more sympathetic to her cause.

In most stories teleportation is viewed as a godsend. But in Stephen King's "The Jaunt" the author explores the implications of what happens if there are dangerous side effects to teleportation. In the future, teleportation is commonplace and fondly called "The Jaunt." Just before teleporting to Mars, a father explains to his children the curious history behind the Jaunt, that it was first discovered by a scientist who used it to teleport mice, but the only mice that survived teleportation were ones that had been anesthetized. Mice that were awake while being teleported died horribly. So humans are routinely put to sleep before they are teleported. The only man who was ever teleported while awake was a convicted criminal who was promised a full pardon if he submitted to this experiment. But after being teleported, he suffered a massive heart attack, uttering the last words, "It's eternity in there."

Unfortunately, the son, hearing this fascinating tale, decides to hold his breath so that he won't be anesthetized. The results are tragic. After being teleported he suddenly goes insane. His hair turns white, his eyes are yellowed with age, and he tries to claw out his eyes. The secret is now revealed. Physical matter is teleported instantly, but to the mind the trip takes an eternity, time appears endless, and the person is driven totally insane.

Page 64: "For the first time," said Eugene Polzik, one of the researchers . . . Curt Suplee, "Top 100 Science Stories of 2006," *Discover Magazine,* December 2006, p. 35.

Page 64: "We're talking about a beam of about 5,000 particles . . ." Zeeya Merali, *New Scientist Magazine,* June 13, 2007.

Page 68: "With luck, and with the help of recent theoretical advances, . . ." David Deutsch, *New Scientist Magazine,* November 18, 2006, p. 69.

5: TELEPATHY

Page 71: The careers of several magicians and mentalists, in fact, have been based . . . At dinner parties one can also perform amazing feats of telepathy. Ask everyone at a party to write down a name on a slip of paper and put the slips in a hat. One by one you pick out a sealed slip of paper and, before opening it, read aloud the name written on it. The audience will be stunned. Telepathy has been demonstrated right before their eyes. Some magicians, in fact, have risen to fame and fortune primarily because of this trick.

(The secret to this amazing feat of mind reading is the following. Pull out the first slip of paper and read it silently to yourself, but announce that you are having difficulty reading it because the "psychic ether" is clouded. Pull out a second slip of paper but don't open it yet. Now recite the name you read on the first slip of paper. The person who wrote that first name will be amazed, thinking you have read the sealed, second slip of paper. Now open up the second slip of paper and silently read it to yourself. Pull out the third sealed slip of paper, and read aloud the name on the second slip of paper. Repeat this process. Each time you say aloud the name on a slip of paper, you are reading the contents of the previous slip of paper.)

Page 72: Gamblers also are able to read people's minds . . . A person's mental state can be roughly determined by tracing the precise path taken by a roving eye as it scans a photograph. By shining a thin light beam onto the eyeball, a reflected image of the beam can be cast onto the wall. By tracing out the path taken by this reflected beam of light on the wall one can then reconstruct precisely where the eye is roving as it scans a picture. (When scanning a person's face in a picture, for example, the observer's eye usually moves rapidly back and forth between the person's eyes in the picture, and then wanders to the mouth, and back to the eyes, before it scans the entire picture.)

As a person scans a picture, one can calculate the size of his pupils and hence whether he experiences pleasurable or unpleasurable thoughts, as it scans particular parts of a picture. In this way, one can read a person's emotional state. (A murderer, for example, would experience strong emotions as he looks at a picture of a murder scene and scans the precise location of the body. Only the murderer and the police would know the location.)

Page 73: The first scientific studies of telepathy . . . The Society for Psychical Research included Lord Rayleigh (Nobel laureate), Sir William Crookes (inventor of the Crookes tube used in electronics), Charles Richet (Nobel laureate), American psychologist William James, and Prime Minister Arthur Balfour. Its supporters have included such luminaries as Mark Twain, Arthur Conan Doyle, Alfred Lord Tennyson, Lewis Carroll, and Carl Jung.

Page 73: One researcher connected with the society... Rhine originally planned to become a minister, but then switched to botany while attending the University of Chicago. After attending a talk in 1922 given by Sir Arthur Conan Coyle, who was giving lectures around the country about communicating with the dead, Rhine became fascinated with psychic phenomena. Later he read the book *The Survival of Man*, by Sir Oliver Lodge, about purported communications with the deceased during séances, which further cemented Rhine's interest. He was, however, dissatisfied with the current state of spiritualism; its reputation was often tarred with unsavory tales of frauds and trickery. In fact, Rhine's own investigations exposed a certain spiritualist, Margery Crandon, as a fraud, earning him the scorn of many spiritualists, including Conan Doyle.

Page 74: "There is left then, only the telepathic explanation..." Randi, p. 51

Page 74: Further tests showed that the mice possessed no telepathic power... Randi, p. 143.

Page 79: In particular, he noticed unusual activity... *San Francisco Chronicle*, November 26, 2001.

Page 80: Some critics also claim... Lastly, there are also legal and moral questions if limited forms of telepathy become commonplace in the future. In many states it is illegal to tape-record a person's phone conversation without his or her permission, so in the future it might be illegal to record one's thought patterns without his or her permission as well. Also civil libertarians may object to reading a person's thought patterns without his or her permission, in any context. Given the slippery nature of a person's thoughts, it may never be legal to enter thought patterns in a court of law. In *Minority Report*, starring Tom Cruise, there was the ethical question of whether you can arrest someone for a crime that the person hasn't committed yet. In the future there might be the question of whether a person's intention to commit a crime, as evidenced by thought patterns, constitutes incriminating evidence against that person. If a person makes threats verbally, would that count as heavily as if a person made these threats mentally?

There will also be the question of governments and security agencies that do not care about any laws whatsoever and subject people involuntarily to brain scans. Would this constitute proper legal behavior? Would it be legal to read the mind of a terrorist to find out his or her plans? Would it be legal to implant false memories in order to deceive individuals? In *Total Recall*, starring Arnold Schwarzenegger, the question arose continually whether a person's memories were real, or implanted, which affects the very nature of who we are.

These questions are likely to remain purely hypothetical for decades to come, but as the technology slowly advances, inevitably the technology will raise moral, legal, and societal issues. Fortunately, we have plenty of time to sort them out.

Page 81: "But if that's the device you want to build . . ." Douglas Fox, *New Scientist Magazine,* May 4, 2006.

Page 81: In this sense, an fMRI mental translator may make it possible . . . Philip Ross, *Scientific American,* September 2003.

Page 83: Such a device would never be as sophisticated . . . Science Daily, www.sciencedaily.com, April 9, 2005.

Page 84: This process is difficult and tedious, since you have to carefully process out spurious waves . . . Cavelos, p. 184.

6: Psychokinesis

Page 90: Before Geller's appearance, Carson consulted with Randi . . . The Amazing Randi, disgusted that professional magicians skilled at fooling gullible individuals could claim psychic powers and hence defraud the unsuspecting public, began a career of exposing fakes. In particular he took delight in duplicating every feat performed by the psychics. The Amazing Randi is in the tradition of the Great Houdini, a magician who also began a second career exposing fakes and charlatans who would use their magical skills to defraud others for private gain. In particular, Randi boasts that he can even deceive scientists with his tricks. He says, "I can go into a lab and fool the rear ends off any group of scientists." Cavelos, p. 220.

Page 93: The National Research Council's report studied creating a hypothetical "First Earth battalion" . . . Cavelos, p. 240.

Page 93: The report concluded that there was "no scientific justification . . ." Cavelos, p. 240.

Page 95: By training them to vary their brain waves . . . Philip Ross, *Scientific American,* September 2003.

Page 95: These monkeys were then able to control . . . Miguel Nicolelis and John Chapin, *Scientific American,* October 2002.

Page 96: "Then I knew that everything could go forward . . ." Kyla Dunn, *Discover Magazine,* December 2006, p. 39.

Page 101: But, he admits, "it still takes the best groups in the world . . ." Aristides A. G. Requicha, "Nanorobots," http://www.lmr.usc.edu/~lmr/publications/nanorobotics.

7: Robots

Page 104: even renowned physicist Roger Penrose of Oxford . . . Professor Penrose argues that quantum effects must be present in the brain that make possible human thought. Most computer scientists would say that each neuron in the brain can be duplicated by a complex series of transistors; hence the brain can be reduced to a classical device. The brain is supremely complicated but in essence consists of a bunch of neurons whose behavior can be duplicated by transistors. Penrose disagrees. He claims that there are structures in a cell, called microtubules, that exhibit quantum behavior, so the brain can never be reduced to a simple collection of electronic components.

Page 104: Colin McGinn of Rutgers University says that artificial intelligence... Kaku, *Visions*, p. 95.

Page 111: Steve Grand, director of the Cyberlife Institute, says... Cavelos, p. 90.

Page 111: "He failed, and I failed on the same problem in my 1981 Ph.D. thesis." Rodney Brooks, *New Scientist Magazine*, November 18, 2006, p. 60.

Page 112: "It doesn't mean Kasparov isn't a deep thinker..." Kaku, *Visions*, p. 61.

Page 113: Not surprisingly, Lenat's motto is, Intelligence is 10 million rules. Kaku, *Visions*, p. 65.

Page 114: "We were killing ourselves trying to create a pale shadow..." Bill Gates, *Skeptic Magazine*, vol. 12, no. 12, 2006, p. 35.

Page 114: "Even something as simple as telling the difference between an open door and a window can be devilishly tricky for a robot." Bill Gates, *Scientific American*, January 2007, p. 63.

Page 115: "No one can say with any certainty when–or if–this industry..." Scientific American, January 2007, p. 58.

Page 118: "There's no machine today that can do that." Susan Kruglinski, "The Top 100 Science Stories of 2006," *Discover Magazine*, p. 16.

Page 118: Hans Moravec says, "Fully intelligent machines will result..." Kaku, *Visions*, p. 76.

Page 119: " 'Please! Please! I need this! It's so important...' " Kaku, *Visions*, p. 92.

Page 120: Neurologist Dr. Antonio Damasio of the University of Iowa... Cavelos, p. 98.

Page 120: "Computers just don't get it." Cavelos, p. 101.

Page 120: As Russian novelist Fyodor Dostoevsky wrote... Barrow, *Theories of Everything*, p. 149.

Page 121: "Our successors will be amazed by the amount of scientific rubbish..." Sydney Brenner, *New Scientist Magazine*, November 18, 2006, p. 35.

Page 124: "It is possible that we may become pets of the computers..." Kaku, *Visions*, p. 135.

Page 124: "When that happens, our DNA will find itself out of a job,..." Kaku, *Visions*, p. 188.

Page 124: So in the long term some have advocated a merging of carbon and silicon technology... So our mechanical creations may ultimately be the key to our long-term survival. As Marvin Minsky says, "We humans are not the end of evolution, so if we can make a machine that's as smart as a person, we can probably also make one that's much smarter. There's no point in making just another person. You want to make one that can do things we can't." Kruglinski, "The 100 Top Science Stories of 2006," p. 18.

Page 125: In the far future, robots or humanlike cyborgs... Immortality, of course, is something that people have desired ever since humans, alone in the animal kingdom, began to contemplate our own mortality. Commenting on immortality, Woody Allen once said, "I don't want to achieve immortality through my work. I want to achieve it through not dying. I don't want to live on in the hearts of my countrymen. I would rather live on in my apartment." Moravec, in particular, believes that in the far future we will merge with our creations to create a higher order of intelligence. This would require duplicating the 100 billion neurons that are in our brain, each of which in turn is connected to perhaps several thousand other neurons. As we sit on the operating room table, there is a robot shell lying next to us. Surgery is performed such that as we remove a single neuron a duplicate silicon neuron is created in the robot shell. As time goes by every single neuron in our body is replaced by a silicon neuron in the robot, so that we are conscious throughout the operation. At the end, our entire brain has been continuously transferred into the robot shell while we witnessed the entire event. One day we are dying in our decrepit, decaying body. The next day we find ourselves inside immortal bodies, with the same memories and personality, without losing consciousness.

8: Extraterrestrials and UFOs

Page 132: Nevertheless, Seth Shostak, senior astronomer at SETI, optimistically believes... Jason Stahl, *Discover Magazine,* "Top 100 Stores of 2006," December 2006, p. 80.

Page 134: "It's hard to imagine how life could survive that extreme onslaught," he says. Cavelos, p. 13.

Page 135: French astronomer Dr. Jacques Lasker estimates that... Cavelos, p. 12.

Page 136: "We believe that life in the form of microbes..." Ward and Brownlee, p. xiv.

Page 137: "We're the first generation that has a realistic chance of discovering life on another planet." Cavelos, p. 26.

Page 146: As I've discussed in my previous books... In general, although local languages and cultures will continue to thrive in different regions of the Earth, there will emerge a planetary language and culture that spans the continents. This global and local culture will exist simultaneously. This situation already exists with regards to the elites of all societies.

There are also forces that oppose this march to a planetary system. These are the terrorists who unconsciously, instinctively, realize that the progression to a planetary civilization is one that will make tolerance and secular pluralism a centerpiece of their emerging culture, and this prospect is a threat to people who feel more comfortable living in the last millennium.

9: Starships

Page 155: Mathematician and philosopher Bertrand Russell once lamented... Kaku, *Hyperspace,* p. 302.

Page 176: Nordley says, "With a constellation of pinhead-sized spacecraft . . ." Gilster, p. 242.

10: Antimatter and Anti-universes

Page 183: Dr. Steven Howe, of Synergistics Technologies in Los Alamos . . . NASA, http://science.nasa.gov, April 12, 1999.

Page 187: He wrote, "It is more important to have beauty in one's equations than to have them fit experiments . . ." Cole, p. 225.

11: Faster than Light

Page 203: As physicist Matt Visser of Washington University says. . . . Cavelos, p. 137.

Page 203: Sir Martin Rees, Royal Astronomer of Great Britain, even says . . . Kaku, *Parallel Worlds,* p. 307.

Page 204: "I thought there should be a way of using these concepts . . ." Cavelos, p. 151.

Page 204: "In back, they wouldn't see anything–just black–because the light of the stars . . ." Cavelos, p. 154.

Page 207: "We would need a series of generators of exotic matter . . ." Cavelos, p. 154.

Page 210: "Pass through this magic ring and–presto! . . ." Kaku, *Parallel Worlds,* p. 121.

Page 211: He says, "You need about minus one Jupiter mass to do the job . . ." Cavelos, p. 145.

Page 211: "But it will also turn out that the technology for making wormholes . . ." Hawking, p. 146.

12: Time Travel

Page 216: In the novel Janus Equation, *writer G. Spruill explored one . . .* Nahin, p. 322.

Page 217: "As for the present, if it were always present and never moved . . ." Pickover, p. 10.

Page 222: "Because we physicists have realized that the nature of time . . ." Nahin, p. ix.

Page 223: As physicist Richard Gott has said, "I don't think there's any question . . ." Pickover, p. 130.

Page 224: Gott says, "A collapsing loop of string large enough . . ." Kaku, *Parallel Worlds,* p. 142.

Page 225: "If he marries in the past can he be tried for bigamy . . ." Nahin, p. 248.

13: Parallel Universes

Page 232: Henderson writes, "Like a Black Hole, . . ." Kaku, *Hyperspace,* p. 22.

Page 233: "At first glance, I like your idea enormously . . ." Pais, p. 330.

Page 235: Enrico Fermi, horrified at the proliferation of subatomic particles... Kaku, *Hyperspace*, p. 118.

Page 239: Max Tegmark of MIT believes that in fifty years... Max Tegmark, *New Scientist Magazine*, November 18, 2006, p. 37.

Page 242: Schrödinger railed against this interpretation of his theory... Cole, p. 222.

Page 242: "So I hope you can accept nature as She is–absurd." Greene, p. 111.

Page 244: Another viewpoint on the paradox is the "many worlds" idea... Yet another attractive feature of the "many worlds" interpretation is that no further assumptions other than the original wave equation are required. In this picture we never have to collapse wave functions or make observations. The wave function simply divides all by itself, automatically, without any intervention or assumptions from the outside. In this sense, the "many worlds" theory is simpler conceptually than all the other theories, which require outside observers, measurements, collapses of waves, and so forth. It is true that we are burdened with infinite numbers of universes, but the wave function keeps track of them, without any further assumptions from the outside.

One way to understand why our physical universe seems so stable and secure is to invoke decoherence, that is, that we have decohered from all these other parallel universes. But decoherence does not eliminate these other parallel universes. Decoherence only explains why our universe, among an infinite set of universes, seems so stable. Decoherence is based on the idea that universes can split into many universes, but that our universe, via interactions from the environment, becomes quite separated from these other universes.

Page 244: Nobel laureate Frank Wilczek says, "We are haunted..." Kaku, *Parallel Worlds*, p. 169.

14: Perpetual Motion Machines

Page 257: "It was Santa Claus and Aladdin's lamp of the whole world," Asimov, p. 12.

Page 265: In theory, a perpetual motion machine of the second type... Some people have objected, declaring that the human brain, representing perhaps the most complex object ever created by mother nature in the solar system, violates the Second Law. The human brain, consisting of over 100 billion neurons, is unrivaled in complexity by anything out to 24 trillion miles of the Earth, to the nearest star. But how can this vast reduction in entropy be compatible with the Second Law, they ask? Evolution itself seems to violate the Second Law. The answer to this is that the decrease in entropy created by the rise of higher organisms, including humans, came at the expense of raising the total entropy elsewhere. The decrease in entropy created by evolution is more than balanced out by the increase in entropy in the surrounding environment, that is, the entropy of sunlight hitting the Earth. The creation of the human brain via evolution does lower entropy, but this is more than compensated for by the chaos that we create (e.g., pollution, waste heat, global warming, etc.).

Page 269: One of the proponents of this idea . . . Tesla, however, was also a tragic figure, probably cheated out of the royalties of many of his patents and inventions that paved the way for the coming of radio, TV, and the telecommunications revolution. (We physicists, however, have guaranteed that the name of Tesla will not be forgotten. We have named the unit of magnetism after him. One tesla equals 10,000 gauss, or roughly twenty thousand times the magnetic field of the Earth.)

Today he is largely forgotten, except that his more eccentric claims have become the stuff of conspiracy buffs and urban legend. Tesla believed that he could communicate with life on Mars, solve Einstein's unfinished unified field theory, split the Earth in half like an apple, and develop a death ray that could destroy ten thousand airplanes from a distance of 250 miles. (The FBI took his claim of a death ray so seriously that it seized much of his notes and laboratory equipment after his death, some of which are still kept in secret storage even today.)

Tesla was at the height of his fame in 1931 when he made the front page of *Time* magazine. He regularly dazzled the public by unleashing huge bolts of lightning, containing millions of volts of electrical energy, to gasping audiences. Tesla's undoing, however, was that he was notoriously sloppy with his finances and his legal affairs. Pitted against the battery of lawyers representing the emerging electrical giants of today, Tesla lost control over his most important patents. He also began to show signs of what is today called OCD (obsessive-compulsive disorder), being obsessed with the number "three." He later became paranoid, living in destitution in the New Yorker Hotel, fearing being poisoned by his enemies, and was always one step ahead of his creditors. He died in total poverty at the age of eighty-six in 1943.

Epilogue: The Future of the Impossible

Page 286: Astronomer John Barrow notes, "Historians still debate . . ." Barrow, *Impossibility*, p. 47.

Page 286: Mathematician David Hilbert, in rejecting Comte's claims . . . Barrow, *Impossibility*, p. 209.

Page 286: "Two hundred years ago, you could ask anybody, . . ." Pickover, p. 192.

Page 287: "All the great questions about the nature of the Universe–from its beginning to its end–turn out to be unanswerable." Barrow, *Impossibility*, p. 250.

Page 287: "But gravitational waves from [the] inflation area are relics of the universe . . ." Rocky Kolb, *New Scientist Magazine*, November 18, 2006, p. 44.

Page 290: "These efforts will reveal intimate details of the Big Bang singularity . . ." Hawking, p. 136.

Page 291: "Do the laws of physics permit highly advanced civilizations . . ." Barrow, *Impossibility*, p. 143.

Page 294: "In 2056, I think you'll be able to buy a T-shirt . . ." Max Tegmark, *New Scientist Magazine*, November 18, 2006, p. 37.

Page 296: Today the leading (and only) candidate for a theory . . . The reason for this is that when we take Einstein's theory of gravity and add quantum corrections, these corrections instead of being small are infinite. Over the years physicists have devised a number of tricks to eliminate these infinite terms, but they all fail for a quantum theory of gravity. But in string theory these corrections vanish exactly for several reasons. First, string theory has a symmetry, called supersymmetry, which cancels many of these divergent terms. Also string theory has a cutoff, the length of string, which helps to control these infinities.

The origin of these infinities actually goes back to classical theory. Newton's inverse-square law says that the force between two particles is infinite if the distance of separation goes to zero. This infinity, which is apparent even in Newton's theory, carries over to the quantum theory. But string theory has a cutoff, the length of the string, or the Planck length, which allows us to control these divergences.

Page 299: "We would then be able to observe them in the sky . . ." Alexander Vilenkin, *New Scientist Magazine*, November 18, 2006, p. 51.

Page 300: Astrophysicist John Barrow summarizes this logic this way . . . Barrow, *Impossibility*, p. 219.

BIBLIOGRAPHY

Adams, Fred, and Greg Laughlin. *The Five Ages of the Universe: Inside the Physics of Eternity.* New York: Free Press, 1999.

Asimov, Isaac. *The Gods Themselves.* New York: Bantam Books, 1990.

Asimov, Isaac, and Jason A. Shulman, eds. *Isaac Asimov's Book of Science and Nature Quotations.* New York: Weidenfeld and Nicholson, 1988.

Barrow, John. *Between Inner Space and Outer Space.* Oxford, England: Oxford University Press, 1999.

——. *Impossibility: The Limits of Science and the Science of Limits.* Oxford, England: Oxford University Press, 1998.

——. *Theories of Everything.* Oxford, England: Oxford University Press, 1991.

Calaprice, Alice, ed. *The Expanded Quotable Einstein.* Princeton, NJ: Princeton University Press, 2000.

Cavelos, Jeanne. *The Science of Star Wars: An Astrophysicist's Independent Examination of Space Travel, Aliens, Planets, and Robots as Portrayed in the Star Wars Films and Books.* New York: St. Martin's Press, 2000.

Clark, Ronald. *Einstein: The Life and Times.* New York: World Publishing, 1971.

Cole, K. C. *Sympathetic Vibrations: Reflections on Physics as a Way of Life.* New York: Bantam Books, 1985.

Crease, R., and C. C. Mann. *Second Creation.* New York: Macmillan, 1986.

Croswell, Ken. *The Universe at Midnight.* New York: Free Press, 2001.

Davies, Paul. *How to Build a Time Machine.* New York: Penguin Books, 2001.

Dyson, Freeman. *Disturbing the Universe.* New York: Harper and Row, 1979.

Ferris, Timothy. *The Whole Shebang: A State-of-the-Universe(s) Report.* New York: Simon and Schuster, 1997.

Folsing, Albrecht. *Albert Einstein.* New York: Penguin Books, 1997.

Gilster, Paul. *Centauri Dreams: Imagining and Planning Interstellar Exploration.* New York: Springer Science, 2004.

Gott, J. Richard. *Time Travel in Einstein's Universe.* Boston: Houghton Mifflin Co., 2001.

Greene, Brian. *The Elegant Universe: Superstrings, Hidden Dimensions, and the Quest for the Ultimate Theory.* New York: W. W. Norton, 1999.

Hawking, Stephen W., Kip S. Thorne, Igor Novikov, Timothy Ferris, and Alan Lightman. *The Future of Spacetime.* New York: W. W. Norton, 2002.

Horgan, John. *The End of Science.* Reading, Mass.: Addison-Wesley, 1996.

Kaku, Michio. *Einstein's Cosmos.* New York: Atlas Books, 2004.

——. *Hyperspace.* New York: Anchor Books, 1994.

——. *Parallel Worlds: A Journey Through Creation, Higher Dimensions, and the Future of the Cosmos.* New York: Doubleday, 2005.

——. *Visions: How Science Will Revolutionize the 21st Century.* New York: Anchor Books, 1997.

Lemonick, Michael. *The Echo of the Big Bang.* Princeton, NJ: Princeton University Press, 2005.

Mallove, Eugene, and Gregory Matloff. *The Starflight Handbook: A Pioneer's Guide to Interstellar Travel.* New York: Wiley and Sons, 1989.

Nahin, Paul J. *Time Machines.* New York: Springer Verlag, 1999.

Pais, A. *Subtle Is the Lord.* New York: Oxford University Press, 1982.

Pickover, Clifford A. *Time: A Traveler's Guide.* New York: Oxford University Press, 1998.

Randi, James. *An Encyclopedia of Claims, Frauds, and Hoaxes of the Occult and Supernatural.* New York: St. Martin's Press, 1995.

Rees, Martin. *Before the Beginning: Our Universe and Others.* Reading, Mass.: Perseus Books, 1997.

Sagan, Carl. *The Cosmic Connection: An Extraterrestrial Perspective.* New York: Anchor Press, 1973.

Thorne, Kip S. *Black Holes and Time Warps: Einstein's Outrageous Legacy.* New York: W. W. Norton, 1994.

Ward, Peter D., and Donald Brownlee. *Rare Earth: Why Complex Life Is Uncommon in the Universe.* New York: Springer Science, 2000.

Weinberg, Steve. *Dreams of a Final Theory: The Search for Fundamental Laws of Nature.* New York: Pantheon Books, 1992.

Wells, H. G. *The Time Machine: An Invention.* London: McFarland and Co., 1996.

INDEX

ALSO BY MICHIO KAKU

BEYOND EINSTEIN

The Cosmic Quest for the Theory of the Universe

Beyond Einstein takes readers on an exciting excursion into the discoveries that have led scientists to superstring theory. What is superstring theory and why is it important? This revolutionary breakthrough may well be the fulfillment of Albert Einstein's lifelong dream of a Theory of Everything, uniting the laws of physics into a single description explaining all the known forces in the universe.

Science/978-0-385-47781-9

VISIONS

How Science Will Revolutionize the 21st Century

In this thrilling tour, Michio Kaku examines the ways the great scientific revolutions of the twentieth century—quantum mechanics, biogenetics, and artificial intelligence—will transform the way we live in the twenty-first century. His unique and compelling vision, based on research already underway at top laboratories around the world, predicts a future in which we are no longer passive bystanders to the dance of the universe, but creative choreographers of matter, life, and intelligence.

Science/978-0-385-48499-2

ALSO AVAILABLE

Hyperspace, 978-0-385-47705-5
Parallel Worlds, 978-1-4000-3372-0